D1423429

Ring Culture

by the same author

*

TOMATOES FOR EVERYONE

Stonor's 'Moneymaker' in the garden, *6th July 1957*
Note that fruits are turning colour

RING CULTURE

FRANK W. ALLERTON

FABER AND FABER

3 Queen Square

London

First published in 1962
by Faber and Faber Limited
3 Queen Square London WC1
New edition 1972
Printed in Great Britain by
Latimer Trend & Co Ltd Plymouth

ISBN 0 571 04751 3

Acknowledgements

The section *Ring Culture Illustrated* (pages 23–28) has been inspired by the wide appreciation of the *Popular Gardening* series of Pocket Guides published over the 1959–60 seasons. I am much indebted to the Editor, Mr. Gordon Forsyth, for permission to use a selection of the line drawings covering ring culture which I prepared for the Guide on tomato growing.

Contents

Illustrations

———————➤➤➤●◄◄◄———————

Introduction

———————>>>►◄◄◄———————

Ring Culture is a method of growing which provides controlled conditions of food and water supply. Once described as halfway between soil-less culture and normal soil growing, it possesses most of the advantages of both with few of the limitations of either. The aptness of this description will in due course become apparent to the reader but, meanwhile, let us look back over the years to the late 1940's when ring culture was evolved and put into practice at the now closed Tilgate Horticultural Research Station in Sussex.

This research centre was developed around the extensive walled-in gardens and lean-to glasshouses of a typically English country seat, Tilgate Manor. The glasshouses had been cropped for many years and for the lack of the means to carry out efficient sterilization, the soil was becoming exhausted and unproductive. Tomatoes grew but poorly, due to early infection with wilt and root-rot diseases; yet, owing to its quick response to feeding, this was a crop ideally suited to the exacting nutritional work then being undertaken.

In view of the fact that equipment for steaming was

not available and soil-changing was reckoned too laborious a project, it was decided to base the work on plants grown in a maiden loam compost contained in some form of pot. Now, it so happened that available to hand was a supply of the large, heavy, glazed earthenware cylinders employed under the name of 'Raschig rings' in the manufacture of sulphuric acid. It was considered that, providing these cylinders were placed in their final positions in the houses prior to being filled with compost, their comparatively large capacity should make them ideal as 'pots' for experiments on tomato nutrition.

In all probability, this use of cylinders or 'rings' would have extended no further than mere expediency if a particular glasshouse had not been used on one occasion for this work. This house, in the manner of the old-time bench-display conservatories, had a tiled floor complete with drainage channels. Part of the floor had at some time been covered with coal-fired boiler ash; the rest was bare tiles. There seemed no purpose in removing the ash so the Raschig rings were set out irrespective of the floor surface, were filled with compost of J.I.P. type and young tomato plants were duly put into the rings. Since fresh maiden loam had been used for preparing the compost, the plants were unhampered by root diseases and the work proceeded under ideal conditions of plant growth.

By half-way through the cropping season, it became apparent, however, that the plants in the rings standing on the ash layer were considerably more vigorous and carried much heavier trusses of fruit than even the excellent plants in the tile-based rings. The final cropping figures fully confirmed the superior appearance of the ash-based plants; these produced an average of 24·4 lb. of ripe fruit per plant compared with 13·1 lb. per plant from the rest. Bearing in mind that all the plants were grown in the same compost and under controlled conditions of nutrition, the marked difference in cropping— from above average on the tiles to quite exceptional on the ash—seemed inexplicable. No answer was found until the rings were duly removed. Then it was apparent that the plants based by ash had made a colossal secondary root system in that layer whereas those on the tiles had, of course, been restricted to the rings. The ash layer, continually moist as a result of overhead damping and ring watering had, in fact, encouraged the plants to extend their roots far beyond their normal spread in soil. These plants had always found an abundance of water and had benefited thereby in no uncertain manner.

Thus did 'ring culture' come into being; the advantages of inducing virtually a double root system were no less obvious than the name by which the system logically should, and did in fact, become generally known in the years that followed.

From the time of that first chance circumstance of cropping, I was privileged to be closely associated with the work on ring culture and was able to follow from season to season over the next ten years or so, the meticulously careful experiments conducted to investigate all possible aspects of this original concept of plant culture.

Many an hour of interesting discussion was spent

with the Superintendent, Mr. H. I. Kingston, and his staff and, as a horticultural journalist in my private capacity, it fell to me to make ring culture known throughout the growing world. One of the first detailed articles on this subject was published in the *Gardeners' Chronicle* and thereafter followed progress reports and step-by-step series in the trade and gardening journals. In particular, that leading weekly, *Popular Gardening*, has been responsible for making the system known to gardeners throughout the country and it is largely as a result of the keen interest evinced by *P.G.* readers of recent years that I have been encouraged to write this book on the specific topic of ring culture.

My earlier book in the Faber horticultural series, *Tomatoes for Everyone*, was based largely on ring culture and much of what I wrote there still applies. Inevitably, however, longer experience of this system has led to improved and more certain methods of application while a constant stream of readers' queries each season has given me a detailed insight of aspects of the system which puzzle newcomers and of troubles that can arise.

In this book, I have attempted to collate in a simple straightforward form, and without assuming more than a rudimentary knowledge of gardening on the part of the reader, the whole subject of ring culture as it has developed and expanded over the years.

Much of this book concerns tomato growing in the greenhouse and out-of-doors. Surprise may be expressed that more space has not been devoted to other crops, particularly as those taking part in B.B.C. Television gardening programmes have from time to time advocated

Introduction

the system for virtually everything from rhubarb to roses. My answer is that ring culture provides particular and quite definite advantages with the tomato mainly by reason of the habit of growth and fruiting of this plant. Most other subjects, though certainly capable of being grown by this system, do at least as well by traditional methods of culture.

Ring culture is certainly the gardeners' answer to tomato growing and I make that claim in the knowledge that it has enabled many thousands to grow this crop effectively where before they looked upon it as scarcely worth greenhouse or garden space. I hope that this book will enthuse many others to adopt this method of growing.

1

The Ring Culture System

Ring culture involves growing a plant in a bottomless pot, cylinder or 'ring' containing a soil-based compost and standing upon a nutritionally inert layer, the 'aggregate'.

From the brief description in the Introduction of the first Tilgate ring plants, it will be remembered that roots are produced in the aggregate as well as in the compost. This rooting in the inert layer upon which the rings are standing is a fundamental feature of ring culture; without such rooting the system falls down completely and, by reason of the labour-saving watering technique peculiar to the system, the plants would be worse off than those limited in rooting capacity to the confines of a pot.

For convenience of description, the aggregate roots are commonly referred to as comprising a 'secondary' root system, those formed in the ring being, of course, the 'primary' system. Although I have found this a useful way of referring to the two apparently separate root systems produced by a ring plant, there is, in fact, no botanical distinction between the two as the roots

B   17

which, under suitable circumstances, grow so prolifically in the aggregate below are merely a downward extension of some of the larger roots which start off from the rootstock of the plant and first grow through the ring compost.

There are, however, considerable differences in function and some differences in structure between the roots which permeate the ring compost and those which extend through the aggregate. Thus the primary function of the ring roots is to absorb the nutrients required to grow leaf and stem and fruit. The aggregate roots, on the other hand, are concerned solely with absorbing water; with supplying the vast quantities of water used by a plant in building up its structure and in breathing out or 'transpiring' vapour from the myriad of pores in the leaf surfaces.

The different functions of these two sets or layers of roots has an influence upon the structure and appearance of the individual roots. Thus we find that the roots formed in the ring compost are closely matted and fibrous whereas those extending through the aggregate are thicker, whiter and usually less branched and matted. These differences are really what one would expect since, with the more familiar example of a plant growing in the open garden soil, the feeding roots in the rich top soil are fibrous whereas those extending down into the poor subsoil in search of deep water are thicker and less branched.

Ring culture is, in fact, nothing more or less than a convenient means of persuading a plant to divide its root system into, functionally, two separate parts; one

18

part concerned mainly with taking up the food elements and the other devoted entirely to water absorption.

Now the true gardener is an enquiring sort of chap, never taking anything for granted, and many who have read my articles and become interested in ring culture have asked, quite reasonably, why this 'divided effort' on the part of the plant should be advantageous. To answer this, we must consider how the roots function with a plant growing in the open soil or in a pot. First and foremost, they have to supply a great deal of water; the amount of water breathed out as vapour by quite a small plant on a hot, dry day is really amazing. Then, of course, the plant itself is composed largely of water— more than nine-tenths of the total weight in the case of the young tender growth—and every stem, leaf, shoot, flower and fruit is charged with water present as sap. The roots themselves are also largely composed of water.

The second and no less important function of the roots is to draw food from the complex mass of materials which go to make up a soil. Roots can, however, only absorb the food required by the plant as a whole when that food is dissolved in the soil moisture. To illustrate this point, a fragment of bone meal applied to the soil in the form of a fertilizer is quite out of reach of the plant until the soil bacteria and minute fungi break down the structure of the bone into simple, water-soluble forms of nitrogen and phosphorus.

The ease with which the roots can absorb water from the soil is dependent to a large extent upon the concentration of food substances in the soil water. The weaker the soil solution, i.e. the nearer it is to plain water, the easier it is for the roots to suck in water and, at the same time, the food materials dissolved in the water. Conversely, the stronger the soil solution in respect of food substances, the more difficult it is for the roots to absorb water and food.

All this means that the more generous we are with fertilizer applications, the more difficult it becomes for the roots to absorb enough water and, if a really heavy excess of fertilizer is applied, the roots, far from being able to absorb more food, will actually shrivel up and die due to water being sucked out of them by the over-strong soil solution.

This brings us back to ring culture in general and to tomato ring culture in particular since, to produce a heavy crop, a comparatively large quantity of food has to be supplied. If this food is applied as a fertilizer to the whole of the root system even in the usual easy stages over the growing season, there is a very real danger that the soil solution will increase in strength to a point where the uptake of water by the roots is restricted. This is particularly likely to happen under glass since, under these artificially hot conditions, the requirements of the plant for water is abnormally high.

Ring culture solves this problem because all the food is applied to the set of roots in the ring and the roots in the aggregate receive only plain water. The result of this 'two-zone' arrangement is that we can feed as heavily as the crop demands in the sure knowledge that the roots in the aggregate, unrestricted in their activity by dissolved nutrients, will be able to supply all the water the plant needs no matter how heavy the crop and how hot the weather.

I have been asked on several occasions how it is that the roots in the ring continue to function in their job of taking up nutrients in view of the quite strong liquid feeding mixture applied to them. Why is their action not suppressed in the manner described where fertilizers are applied to the root system as a whole? The answer, it would seem, is that roots can gradually become adapted to the uptake of nutrient solutions considerably higher than normally exist in soil but that the rate of uptake is slower than when the solution strength is low. This means that these roots would be incapable of supplying all the water requirements of the plant but this, of course, does not matter with ring culture since the aggregate roots are supplying, unhampered, most of the water needed.

This explanation of the working of the two-zone root system might well be criticized as very much over-simplified but I have found it most useful in accounting for the observed facts. It also shows why a strong, healthy secondary root system in the aggregate is so important and why the aggregate roots should never be fed. To do so would obviously defeat the object of un-restricted water uptake.

Now, although the 'whys and wherefores' of the way in which ring culture functions may not be agreed by all concerned there is universal recognition in these days that the system 'works' and does so in no uncertain manner in the case of tomatoes.

Ring Culture Illustrated

A PICTORIAL GUIDE
FOR THE BEGINNER

Cutting Plan for preparing Rings from Roofing Felt

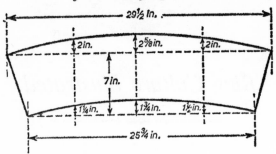

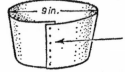

Allow 1–1½ in. overlap and secure with six push-through paper clips

(See page 33)

An alternative to Ash for the Aggregate Layer

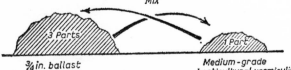

Mix

3 Parts

¾ in. ballast

1 Part

Medium-grade horticultural vermiculite

 or:

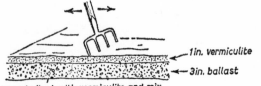

1 in. vermiculite

3 in. ballast

Cover ballast with vermiculite and mix roughly by back and forth raking action with fork

(See page 41)

24

The Aggregate Layer

(See page 43)

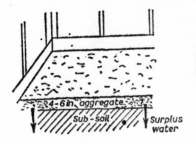

1. Usual aggregate arrangement. The layer preferably covers whole floor of greenhouse but may be restricted to 15–18 in. side and end borders.

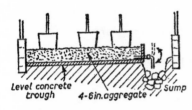

2. A refined method of aggregate watering. The aggregate medium is laid in a concrete trough provided with a pipe outlet which can be swivelled upwards to retain the water while the rings are absorbing by suction, and then downwards to drain off surplus into a sump.

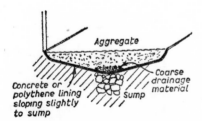

3. An effective means of isolating the aggregate layer from glasshouse soil known to be disease-infected.

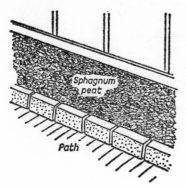

4. With this arrangement, surplus water from the peat escapes across the path through the row of un-cemented bricks.

Planting Arrangements

(See page 57)

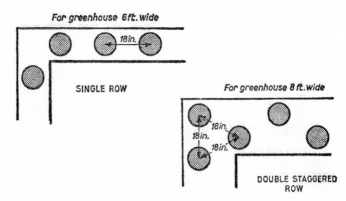

For greenhouse 6ft. wide

18in.

SINGLE ROW

For greenhouse 8ft. wide

18in.
18in.
18in.

DOUBLE STAGGERED ROW

26

Planting in the Rings
(See page 57)

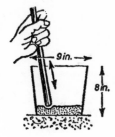

Hold ring firm and ram 2–3 in. of moist compost firmly on to aggregate.	Lightly ram further compost round pot ball taking care not to damage roots.	Adjust pot ball of roots so that, when surface just covered, about 1½ in. left for watering.

Watering and Feeding
(See page 59)

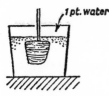

Planting stage. Water to settle roots and compost and to provide moisture reserve.

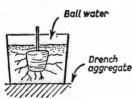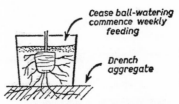

7–14 day stage. Commence ball-watering and drenching aggregate.

4 weeks stage. Roots extensive in aggregate and drawing most of water requirements of plant. Nutrient reserve in ring compost diminishing so commence feeding.

27

Out-door Ring Culture

(See page 77)

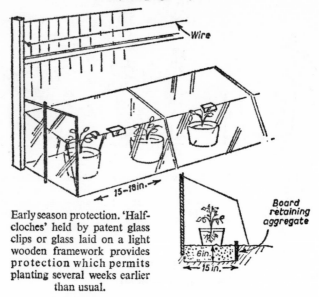

Early season protection. 'Half-cloches' held by patent glass clips or glass laid on a light wooden framework provides protection which permits planting several weeks earlier than usual.

Method of Roof Training

(See page 89)

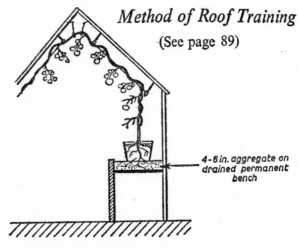

Method of roof training with leading growth of plant taken over and down opposite roof slope. (Leaves omitted for clarity.)

2

The Rings

————————▸▸▸━◗◉◖◀◂◂————————

When ring culture came into being at Tilgate, the rings involved cumbersome glazed porcelain cylinders. These held no less than 24 lb. of J.I.P. compost and this considerable weight, taken in conjunction with the heavy rings, encouraged early experiments to discover whether smaller, lighter rings, perhaps even something of an expendable nature, could serve an equal and more convenient purpose.

The first trials were with 10-in. top diameter clay and Guernsey-type cement pots in which the bottoms had been knocked out to produce 'cylinders' with sloping sides. At the same time, large pots complete with bottoms were tried out to determine whether there was any necessity for the compost to be in actual contact with the aggregate or whether roots escaping through the drainage hole would give rise to equally good results.

Careful crop records for a number of plants over a full season showed that the 24 lb. of J.I.P. type compost in Raschig rings gave no better, earlier or heavier crop than the 17 lb. or so contained in the adapted clay or cement rings. This was a move in the desired direction of making the rings more easily handled and of reducing the requirement of compost but, even so, clays or cement rings were still reckoned too cumbersome and they had the further disadvantages of being easily broken and, especially the cement type, so porous as to cause unduly rapid drying of the compost.

The next step was to compare the bottomless 10-in. clays with rings made from the roofing felt type of flexible composition. These had top and bottom diameters of 9 in. and 8 in. respectively so they provided a considerably greater contact area of the compost with the aggregate than was the case with the clay rings. These flexible rings were of negligible weight and were an adaptation of composition pots then becoming popular commercially. They held 14 lb. of compost compared with 17 lb. in the clays and were sufficiently cheap to be discarded after a season's use.

Cropping results were again virtually identical; in both instances, 18 lb. per plant of ripe fruit being produced in a heated house over the cropping period of May 19th–September 30th.

Since scarcely more than half as much compost as that contained in the Raschig rings had thus been proved adequate, it was decided, purely as a matter of interest, to carry out a small trial to determine whether even smaller ring sizes would suffice. It was found that over a cropping period of mid-June to late September, the 14 lb. capacity rings produced an average 13 lb. of fruit; 12 lb. capacity rings 12 lb. of fruit; $4\frac{1}{2}$ lb. capacity rings 9 lb. of fruit; and $2\frac{1}{2}$ lb. capacity rings $7\frac{1}{2}$ lb. of fruit.

With a decreasing weight of compost, the rate of

cropping was becoming less but to a much smaller degree than had been anticipated. Though nothing smaller than the standard 9-in. rings with their 14 lb. capacity could be justified, even a ring comparable in capacity to an ordinary 48 size pot would produce more than half as much fruit!

From these careful experiments, it was very apparent that in ring culture, as compared with ordinary confined pot culture, the compost was not the only factor determining the crop; that, in fact, the aggregate layer was of parallel importance.

The other experiment, carried out at the same time, to find out whether a ring was any better than a pot, in both cases standing on an aggregate layer, gave equally interesting results.

With rings, the compost made immediate contact with the aggregate whereas with a pot, the bottom, slight as it might be in a composition pot, introduced a barrier between compost and aggregate.

In both cases, a strong root system formed in the aggregate though development was slower where the roots had to escape from the drainage hole or holes of the pot. A big difference was found, however, in the way the compost dried out in the pots as compared with the rings. The pot compost dried out almost as rapidly as if the plants had produced no aggregate roots whereas the ring compost remained moist for much longer periods.

A requirement of ring culture is that the aggregate layer shall be kept thoroughly moist. It was shown that the compost in contact with the moist aggregate was sucking up water whereas this could not, of course, occur where the bottom of the pot provided a barrier between compost and aggregate.

The plants in the pots were much inferior to those identically treated in the rings so there was no longer any doubt that, for effective ring culture, there must be intimate contact between compost and aggregate and that the compost must be of a type capable of sucking up water freely.

The practical significance of these conclusions will be readily apparent in later chapters.

Today, we use 9-in. top diameter flexible composition rings for ring culture whether in the greenhouse or out-of-doors but the foregoing notes on the early Tilgate work will show that what we have come to accept as the obviously correct thing was not originally so certain or obvious!

These rings, usually red in colour, are today freely available from garden shops at around 30p a dozen. Although they often seem quite sound when the crop is finished, I find that breakdown invariably occurs before the completion of a second season. They should be reckoned as expendable and at this price one can surely afford to throw them away. As I have said, composition rings are today freely available because, with thousands of gardeners throughout the country employing ring culture, there is an insistent demand for them during the early months of the year. Nevertheless, we gardeners are the original 'do it yourself' types, so it is understandable that many enthusiasts like to make up their own rings in the winter when the work-bench holds more attraction than the rain-soaked garden.

1. A commercial ring culture crop.

2. *Left:* Root system where ring stood on cinder.
   Centre: Root system where ring stood on unsterilized soil.
   Note that only a few corky roots could be disengaged from the
   soil.
   Right: Root system of plant grown in the unsterilized soil.

The Rings

Rings can, in fact, be made up quite simply from odd strips of any desired grade of roofing felt. The simplest procedure is to open up a manufactured ring by careful removal of the 'stitches', to lay the flattened sheet over the material to be cut and to make the cutting lines with a thick pencil or with chalk. Scissors, preferably (for the sake of household peace) not the jealously guarded embroidery edition, will cut this sort of material without too much difficulty.

Linoleum can be used in the place of roofing felt but the cut-out strips are too stiff to bend into the ring shape without cracking unless first well warmed.

Another alternative is to use sleeves cut from 9-in. diameter, 5-thousandths thickness polythene tubular film. Filling is, however, an awkward job with this floppy material.

I have seen the clear grade of polythene used without apparent objection by the plants but green slime grows on the inside surface and this is known to be deleterious to the roots. For preference, use black polythene as it has the further advantage of resisting breakdown in sunlight and, consequently, the rings can be used for several seasons.

Avoid cardboard in any shape or form as, though strong and rigid when dry, this material loses all strength when in contact with wet soil.

The materials can all be cut out with the help of a pattern but, failing a specimen ring to hand, the cutting can follow the diagram in the 'Ring Culture Illustrated' section.

There is much to be said in favour of expendable

rings; they are light and clean to handle and, serving one crop only, cannot carry root diseases over from one season to another. On the other hand, fine tomato crops have been grown in rings contrived from canteen-sized fruit tins and in bottomless wooden boxes of 7-in. sides. With canteen tins, it is only a matter of removing the bottom with the modern type of clean-cutting tin-opener and, although the capacity for compost is appreciably less than that of a proprietary composition ring and the eventually rusted surface far from sightly, such tins make useful rings.

Open-ended boxes constructed from light timber, such as the boards from margarine or soap cases, can take the place of the conventional rings. One is naturally inclined to treat such containers with a wood preservative but it is advisable not to do so as most preservatives, and especially creosote, are toxic to roots. Better to allow the wood to rot over a couple of season than to jeopardize the crop!

As will be seen, there are several ways in which ring culture containers can be contrived and there is little to choose between the various types, providing the capacity is of the order of 14 lb. or one-fifth bushel of compost and the bottom diameter is at least 6 in. The point about this latter requirement is that sharply tapered containers, apart from being unstable, limit unduly the area of contact between compost and aggregate and so restrict the uptake of water by suction.

3

The Aggregate Layer

According to the dictionary, an aggregate is a collection of mineral particles or pieces formed into a mass or layer.

It is appropriate, therefore, to refer to the layer upon which the ring culture containers stand as 'the aggregate' since we employ just such a collection of mineral fragments.

The first ring culture plants at Tilgate had a layer of weathered greenhouse boiler ash as aggregate and, despite the numerous investigations into alternative materials since that time, boiler ash or, more generally, cinder ash, is still our best and cheapest aggregate material when it is available in suitable form.

Indeed, the only justification for seeking alternatives is that the ash from solid fuel fires is an extremely variable material and one which quite commonly contains substances harmful to plant life. These latter can usually be removed only by prolonged weathering, a circumstance which is annoying for those who have in mind to use winter-collected domestic ash for a ring culture installation in the following spring.

The Aggregate Layer

The residue from domestic fires tends, even with the open grate, to contain an unduly high proportion of dust and soft, flaky particles. This failing of a conveniently 'home-made' waste material has been still further emphasized by the introduction of slow-burning grates and the smokeless fuels. Thus, for example, Coalite which, in any case, burns to a predominantly soft ash, leaves virtually no coarse gritty particles when consumed in the latter type of grate.

Such ash should never be used for ring culture aggregate as it packs down into an airless, sticky mass when watered.

So far as domestic ash is concerned, we have the ideal material in a heap of open-grate coal cinder ash which has stood in some odd corner of the garden for a year or more and where the absence of toxic residues is indicated by the growth of occasional weeds and clumps of grass. With the weeds removed and the surface skimmed off to exclude wind-borne weed seeds, this ash can be filled direct into the ring culture site with every confidence that it will do an excellent job for many years.

I am often asked whether there is any means of rendering winter-collected ash fit for use for the next year's crop. There is unfortunately always a risk that toxic principles will persist but, if the ash has the desirable coarse gritty texture with a good deal of cinder and unburned fuel, it is worth the bother of careful preparation. As the ash is produced, it should be tipped in a layer no more than 3 or 4 in. deep along a little-used path. There it will be washed through frequently by rain, and the sulphur compounds, the main cause of trouble, will become oxidized by the air and gradually leached away. It is as well to give at least three months' weathering so the last date of collecting for the following season is around Christmas. I have personally used coal cinder ash prepared in this manner quite successfully but a reluctance for the plants to root into the aggregate on occasion and the typical toxicity symptom of scorched leaf edges have more than once warned me that the ash had been insufficiently weathered and leached.

One can, of course, make up for the inadequate washing through which results from a dry winter by heavy and repeated soaking of the aggregate layer once the ash is *in situ*. This procedure is, in fact, sound practice however long the ash has been weathered and I always adopt it as a safety measure even when only adding a shallow 'make-up' layer to an existing aggregate.

Failing a supply of suitable domestic ash, it is sometimes possible to obtain a fine grade of clinker from a local coal-fired factory. This is excellent material for the aggregate, as it is virtually free of soft ash and has a honeycomb structure which makes for free aeration. The ideal grade is around $\frac{1}{2}$-in. to dust with the majority of the particles about $\frac{1}{4}$-in. in size. With such clinker, the content of toxic materials is much less than in soft ash and weathering for a month or two is usually quite sufficient to render it safe for use.

Small graded clinker of this type is used considerably in the breeze block industry and it is often possible to persuade cartage contractors to drop off relatively small amounts in the course of bulk deliveries. Complete replacement of the soil of a greenhouse 8 ft. by 6 ft. to a depth of 6 in. requires just on 1 cubic yd. of aggregate material. If it is decided on the grounds of minimizing labour and cost to retain a central earth path, this quantity is reduced by about one-third.

If weathered cinder ash or graded clinker is available, there is no reason to look farther for the means to provide the aggregate but many gardeners have no easy access to the industrial material and do not produce suitable ash from the home consumption of fuel. Substitutes, or rather alternatives, then become of interest and, over the years, I have made a point of investigating several such materials.

Starting from the beginning, the obvious way to search for an alternative to a particular material is to define the properties of that material and then see what else matches up to its properties. What does this involve in the case of cinder ash or fine industrial clinker?

First of all, cinder ash-cum-clinker is a waste product and, as such, without cost when home-produced or quite cheap to acquire from nearby industrial sources. Secondly, it is porous in nature and so can absorb a great deal of water without becoming airless and waterlogged.

Thirdly, it does not change physically in use and so can be used season after season.

Finally, it is to all intents and purposes free, when well weathered, of soluble materials, and consequently, satisfies the previously explained requirement in regard to the unrestricted uptake of water by the secondary root system.

38

The Aggregate Layer

With these properties in mind, we can consider the relative merits of generally available alternatives.

Gravel? Very well aerated, unchanging in condition, easily obtained, not unreasonably expensive, free of soluble materials; all of these properties, certainly, but with one big disadvantage: the stone fragments of which it is composed are not porous so the water-holding capacity is limited to the 'skin' of water around each fragment. Various grades have been tried out several times right from the early Tilgate days but even the fine grades require watering at least twice a day in hot weather and for this reason alone it cannot be recommended.

Sand? Fine sand holds a fair amount of water, but inclines to water-logging and hence poor aeration; coarse sand is little better than fine gravel.

Crushed brick? Quite good, probably as good as cinder if grading from $\frac{1}{4}$-in. to dust and very well weathered but freshly crushed brick or brick kiln fragments seem invariably to have a dangerously high soluble content. Laborious to produce in graded form from old bricks and therefore not usually worth serious consideration.

Ballast? Combines the best properties of gravel and graded sand and constitutes the best straight alternative to cinder-clinker. The $\frac{3}{4}$-in. grade seems best and has many times given me acceptable results.

Vermiculite? Much more expensive than even clinker bought-in from a distance but otherwise at first sight the ideal alternative. Unfortunately, the individual columnar particles of which it is composed collapse beneath the weight of the compost-filled rings which results in a wet, soggy, airless mass under the rings. Not suitable for use on its own.

Peat? Not, it might be thought, at all suitable owing to its organic nature. In fact, however, the best grades of sphagnum peat have such a suitable structure for retaining great quantities of water without becoming airless that a comparatively shallow layer of such peat provides an excellent aggregate. The springy nature of this material prevents undue compression beneath the rings and I have used a 2- to 3-in. layer of peat on numerous occasions with excellent results.

There is one not-too-serious disadvantage, however. The whole cellular structure becomes full of roots which cannot be disentangled at the end of the season. A fresh layer must be put down each season but the used material need not be wasted as the texture has scarcely changed and it can be employed in the garden for digging-in or for mulching. Rather than cover the whole floor of the greenhouse, a procedure which has much to recommend it with mineral aggregates, I find it a better and much cheaper proposition to retain an above-path layer of suitable width by means of a row of bricks laid side-on. With this arrangement surplus water escapes between the bricks whenever the layer is drenched and water-logging is thus prevented.

Since restriction of the roots to the rings for the first fourteen days or so is desirable in the interests of producing a sturdy, fruitful plant, I prefer to make up the aggregate with the peat in the very dry condition that it comes from the bale. Aggregate roots are then not encouraged to form until repeated sprinkling at the appropriate stage gradually charges the peat with water.

Another aggregate medium which I use considerably in the absence of suitable cinder ash or clinker is a mixture of the specified ballast and medium grade horticultural vermiculite. A suitable proportion is about three parts by volume of ballast to one part of vermiculite but, to avoid the heavy job of mixing the two, I put down about 3 in. of ballast, apply an inch layer of vermiculite and work the two materials in together with a fork used as a rake. If the whole floor is covered with the ballast, economy in vermiculite can be achieved by applying this material only to the 18 in. or so wide strip around the walls as this latter constitutes the main secondary rooting area of the plants.

The purpose of the vermiculite is to increase the water-holding capacity of the ballast while the latter protects the little vermiculite columns against compression by the rings. One point of importance: always employ the horticultural, i.e. the slightly acid grade of vermiculite in any project to do with plants; the cheaper insulating grade is usually strongly alkaline and, as such, harmful to roots.

It will be seen that we have two satisfactory alternatives to fuel residues for preparing the aggregate layer. There is no need to risk the dire effects of insufficiently weathered ash and, where there is any doubt, I would strongly advise either sphagnum peat or ballast-vermiculite. There seems little to choose in efficiency between the two.

Out of doors, with the crop to be grown against a

41

south-facing wall or fence, the aggregate is usually laid in a trench cut alongside, to a depth of about 6 in. and a width of 12–15 in. Desirably, the surface of the aggregate should be $\frac{1}{2}$ in. or so above the adjacent border since this minimizes the tendency for soil to wash into the aggregate during the course of watering or in periods of heavy rain (or see diagram p. 28).

On occasion, a hard path rather than a border runs alongside a suitable fence or wall. Under these circumstances, the thing to do is to prepare an above-path level aggregate bed. By erecting a loose wall of cement or breeze blocks, the aggregate can be satisfactorily contained while at the same time allowing free passage of surplus water between the uncemented blocks. Where a fence is involved, it is as well, by the way, to protect the timber from the rotting effects of the continually moist aggregate by means of tiles or asbestos sheet.

Any of the aggregate materials discussed can be used to equally good effect out of doors as in the greenhouse. For a raised aggregate bed, however, I favour sphagnum peat in view of its high retention of water and the limited depth which will suffice with peat; bricks turned side-on will retain an adequate layer and this is certainly a more eye-pleasing arrangement than large blocks.

This brings us logically to the question of the required depth of the aggregate. This was one of the first aspects of ring culture to be investigated at Tilgate. It was found that 4 in. was sufficient with any of the usual materials and that, although there might be slight advantage with a little greater depth, nothing was gained beyond 6 in. As a result, we commonly employ an aggregate layer 4 to 6 in. deep except in the case of sphagnum peat where 3 in. is, I find, quite sufficient. There is, however, no objection to layers of greater depth and such might even prove desirable where the drainage of the soil beneath is very poor.

On the consideration that the aggregate has to be kept continually moist during the main growth period of the plants, I am often asked whether it would not be desirable to check drainage of water from the aggregate by an underlying layer of cement or polythene. This, in fact, is not normally desirable as any tendency towards water-logging must be avoided in the interests of sustained healthy root action.

There is, however, a case for placing an impervious layer between the aggregate and the underlying or surrounding soil where that soil is known to be heavily infected with eelworm or with Verticillium wilt. A suitable scheme is shown in the diagram on page 25 and it will be noted that provision is made for free escape of surplus water.

Passing finally to the positions in which an aggregate may be laid, we have the established ones of the greenhouse floor and the trough or raised bed alongside a wall or fence. A less obvious position is upon built-in benches in the greenhouse. Old-time structures commonly had quite massive metal benches intended to retain a shallow layer of shingle or cinder ash upon which pot plants could be stood. By increasing the effective depth of such benches with a surround of 3–4 in. boards, a ring culture aggregate can be laid.

The only disadvantage of this arrangement is that of

limited headroom for a tall-growing plant such as the tomato but roof training can be practised and details of this will be found in Chapter 8.

Then, again, many town-dwellers have nothing more encouraging than a concreted yard or a flat roof for a garden. Providing the spring and summer sun reaches such positions for several hours on clear days, tomatoes and some other plants can be grown effectively by ring culture with the aggregate laid direct on the concrete and supported by a surround of bricks, boards or blocks as described for a wall-side path. It is, in fact, under such circumstances, where all soil has to be imported, that ring culture is most acceptable in outdoor growing. Cinder ash, ballast or sphagnum peat are all usually easier to acquire than good soil, so ring culture, with its minimum soil requirements, makes yard and roof gardening a practical proposition for those with limited access to the normal requirements of plant culture.

To summarize briefly: Avoid making up the aggregate with gravel, stone chippings or coarse clinker since the water-holding capacity of such materials is limited and the ring compost will dry out unduly. On the other hand, anything as fine as builders' sand, though retentive of moisture, becomes compact to the extent of excluding air to the detriment of the roots trying to penetrate the aggregate.

As a general rule, avoid mixing gravel or other mineral fragments with peat for an aggregate since, when the aggregate is turned out at the end of the season—as, indeed, it must be where a decomposable substance such as peat is involved—the mixture is less acceptable than peat alone for mulching or digging-in. Apart from this aspect there is, however, no objection aggregate-wise to peat/mineral fragment mixtures.

TREATMENT OF EXISTING AGGREGATE BEDS

It is commonly queried whether the aggregate should be subjected to any form of sterilization between crops.

At Tilgate, in the first house laid down to ring culture, the original cinder aggregate had grown nine successive crops of tomatoes when the Station closed, all excellent and with no more sterilizing treatment than that resulting from formaldehyde solution splashed down during the course of spraying the superstructure of the house.

It would seem, then, that unless root diseases are introduced by bought-in plants, there is no necessity to sterilize the aggregate. There is little doubt, however, that the aggregate is capable of carrying infection from one crop to the next; so, if plants fail due to a wilt or eel-worm infection, or the aggregate roots are brown and corky when the crop is cleared, it would be wise to carry out normal formaldehyde treatment as soon as the house is free of plants. Two gallons per square yard of the usual 1 in 50 mixture will be found adequate for this purpose in view of the very open nature of the material. In the comparatively rare instances of eelworm infection, formaldehyde should be replaced by a tar-acid sterilizer.

4

The Ring Compost

————————▸▸▸➤◑➤◅◂◂————————

All the early work on this system involved the use of John Innes Potting Compost No. 3, i.e. J.I.P.3, in the rings.

This was an obvious choice as the Raschig rings of the original crop had been filled with this compost and it had long since become recognized as the standard soil mixture for all forms of cropping containers.

The authentic J.I.P. compost is based upon a medium heavy, 'greasy' loam of high moisture-retaining capacity, and where such loam has been used in the preparation of the ring compost there has been no problem in keeping the compost moist between the weekly feeds (see Chapter 6) by suction of water from the continually moist aggregate beneath. This ideally textured loam, usually identified as 'Cranleigh' or 'Kettering' type, is, however, by no means generally available and most turf loams or garden soils are far too light in texture and sharply drained to produce a good J.I.P. compost even for ordinary pot work. When such compost is used in the rings, it dries out to a very marked degree between feeds with the result that the feeding roots are unable to re-main active and unbalanced growth results.

Thus, if a strong aggregate root system has not been formed, the plants wilt heavily, commence to drop the blossoms on the upper trusses and develop a fruit con-dition known as 'Blossom-end rot' (see Chapter 10). Where, on the other hand, there is a sufficient root sys-tem in the aggregate to satisfy the demand for water, the plant virtually lives, between feeds, on water and be-comes coarse, sappy and generally unbalanced in growth. The fruits are large, soft, watery and lacking in flavour and the condition known as 'blotchy ripening' (see Chapter 10) is rife.

Bound up with this same matter is the fact that a ring compost which is too open and sharply drained to keep itself moist by suction of water from beneath, is also unable to absorb all the liquid feed applied at weekly intervals. This again leads to a degree of under-nourish-ment and of drainage of valuable nutrients into the aggregate beneath, where, when the aggregate is next watered, they are washed away and lost to the plant.

It will be seen, then, that it is vitally important in ring culture to employ a really retentive compost. How can such a compost be prepared when nothing better than a light sandy or chalky loam is available?

In the first place, the compost can be made much more compact in texture and retentive of water by omitting part or all of the sand employed in J.I.P. com-post. Thus, for example, instead of the usual 7 parts by volume loam, 3 parts sphagnum peat and 2 parts coarse sharp sand, one might make the mixture 8 loam, 3 peat and 1 sand, or even, with a very light loam, 8 loam and 4 peat. Good peat has an excellent absorbing and suction power for water but it cannot be used alone in the rings since soil is required to provide trace elements and as a 'reservoir' for nutrients upon which the roots can draw gradually.

My own garden soil, from which I make up the ring compost, is a fine sandy loam and with this as a basis I omit the J.I. grade sand entirely. Even then a compost made up of 8 parts of loam and 4 parts peat is less retentive than experience has led me to consider desirable so, rather than go to the bother and expense of finding and buying-in a heavy loam, I set about improving the texture of what is available to hand.

The texture and water-holding properties of a soil are dependent upon the proportions of clay, silt, sand, humus and so on. Clay has the most marked effect; the lower the clay content, the lighter and less retentive a soil while, conversely, the higher the proportion of clay, the stickier and more retentive it is. An obvious approach, then, is to increase the proportion of clay in the garden soil or turf loam from which one intends to prepare the compost. This is simple enough to do in any desired proportion with finely crushed dry clay but the problem is to obtain the clay and to break it up into particles of size $\frac{1}{4}$ in. or so to dust.

As far as I know, broken clay cannot be bought as such and in many areas of the country there are no clay formations for many miles. However, the advantages to be gained from adding clay to a light soil are so marked that it is well worth making a car trip or persuading a

3. The composition ring has been stripped off to
show both ring and aggregate roots.

4. Roof crop of 'Moneymaker', *August 1961*.

5. A promising crop, *early June*.

friend to bring along a large boxful during the course of a visit.

Then comes the job of breaking down the putty-like lumps into something which can be mixed evenly with the soil. One way is to slice the lumps with a spade and leave the mass to dry out on the floor or bench of a shed or greenhouse. After some weeks, a little vigorous work with a mallet and a $\frac{1}{4}$-in. sieve will achieve the desired effect. Another method which I devised some years ago and which appeals to me since it satisfies my rooted objection for doing unnecessary work, makes use of the winter frosts to break up the lumps. When clay freezes, the contained moisture turns to ice; on thawing, the structure is found to have changed from putty-like to friable and almost granular. We make use of this physical change when we leave heavy soil roughly dug for the frosts to penetrate, and every gardener concerned with such soil knows that this is the only way to achieve a spring tilth.

The best way, I find, to break down sticky clay for ring culture is to spread it, in roughly chopped condition, over a sheet of asbestos or corrugated iron and to protect against rain with another similar sheet supported 3 in. or so clear with corner bricks and duly secured against wind displacement. If this is done in the autumn and even the odd sharp frost is experienced, the clay will be found to have broken down to a near-dry granular layer by the time it is wanted. There only remains then to sieve out the odd stone and break down the few remaining lumps. Definitely a lazy man's way but, as usual, Dame Nature does the job better than we can!

The Ring Compost

Is the type of clay important? One instinctively goes for the warm brown type and, desirably, it should be taken from a foot or two below the surface to ensure virtual freedom from weed seeds, pests and disease organisms. In fact, however, any type of clay, irrespective of colour, seems suitable but make sure that it *is* clay and not merely a sticky silty sand of the type which often overlies gravel beds.

What proportion of the broken-down clay should be mixed into a light soil or turf loam? It depends, of course on how light is 'light'. As a rule, however, it is better in ring culture to err on the heavy rather than on the sandy side and, on this consideration, a mixture of one part by volume of clay to three parts soil will suit most circumstances.

It might be felt that a no-sand, clay-boosted compost would be poorly drained to the extent of becoming water-logged. This might indeed happen, despite the 'sponge' effect of the peat, in a clay pot but it is virtually impossible with a bottomless container standing on a freely-drained aggregate.

Irrespective of this texture factor, the best soil to use for any form of potting and, equally as a basis for ring culture compost, is a loam resulting from the decay of 4–6 in. turves stacked for 6 to 12 months. Few of us, however, have this sort of material available—I seldom have myself—and are therefore forced back on to something much more to hand.

I use the top soil from an area of couch grass-ridden neglected land with roots carefully shaken out. Few weeds grow in the rings and the healthy appearance of

The Ring Compost

the ring roots even right at the end of the season denies the need for sterilizing the soil prior to use. Failing such a source of virtually maiden soil one could use the top inch or two skimmed off a fertile area of allotment or the home vegetable garden. Under these circumstances, it is as well to avoid taking the soil from any area which has grown outdoor tomatoes or potatoes for several seasons as root pests and diseases which could affect the ring culture plants are likely to persist for a long time. With this precaution and providing it is accepted that the almost inevitable presence of some weed seeds will necessitate weeding of the rings from time to time, there is, in my experience, no advantage to be gained by sterilizing the soil. Indeed, sterilization, especially by means of heat, so greatly increases the availability of nitrogen in the soil that early season growth in a compost based on such treated soil tends to be soft, over-vigorous and unfruitful.

Where, however, it is desired for some reason to carry out a form of sterilization, I would advise the avoidance of chemicals such as formaldehyde or the 'carbolic' disinfectant type since weed seeds are not killed and the toxic residues take an inconveniently long time to disperse. Heat treatment is much to be preferred by reason of its thoroughness and one can either use an electric soil sterilizer or carry out controlled baking on a metal sheet. Various types of sterilizing equipment using electricity as a source of heat are available and all are both safe and effective if used strictly according to instructions.

I can particularly recommend the range produced by

the General Electric Co. Ltd., Magnet House, Kingsway, London, W.C.2.

For those who do not wish to incur an outlay in the region of £20 for such a sterilizer, I suggest heating the soil on a metal sheet such as a piece of corrugated iron. My own procedure is to construct a fireplace about 18 in. wide and 3 ft. long with concrete blocks to give support to the metal sheet which will carry the soil. Prunings and other hard garden rubbish are burned in the fireplace until there is a good bed of glowing embers.

The metal sheet is then laid over, more fuel put on the fire, and the soil to be sterilized, previously passed through a $\frac{1}{2}$-in. sieve, is poured on to the sheet to a depth not greater than 2 in.

The soil should now be thoroughly soaked through a fine rose since the purpose is to produce steam rather than to burn the soil. Sacking laid over the wet soil will help to retain the heat and after 10 min. or so clouds of steam will be rising through the sacking.

After a further 5 min., sterilization will be complete and the soil can be swept off on to a clean groundsheet for transfer to a clean concrete surface to cool.

Soil can be sterilized in this way at leisure and can be stored against requirements in a loosely covered box or metal bin. It may require to be moistened prior to use.

I find that by maintaining a brisk fire, having the lump-free soil thoroughly wet and keeping the layer shallow, one can make a thorough job with very little burning of the soil.

The best proof of effective treatment is the virtual absence of weed seedlings when compost is made up with the baked soil since, if weed seeds are destroyed, it is safe to assume that pests, fungus diseases and harmful organisms in general have been eliminated.

So, whether a relatively weed-free turf loam or maiden soil is being used without sterilization or a garden soil after heat treatment, the ring compost becomes:

> by volume, 6 parts moist soil
> 2 parts finely broken dry clay, and
> 4 parts soaked sphagnum peat.

If, as is usually convenient, a 2-gal. bucket is taken as the unit of volume measurement for the above quantities, the 24 gal. or 3 bus. of compost prepared is sufficient to supply 15–16 of the standard 9-in. top diameter, 8 in. deep flexible composition rings.

Peat, by the way, should always be measured as 'loose volume', i.e. without being pressed down in the bucket, and in thoroughly moist condition. Good sphagnum peat should be dust-dry and in this economical condition it takes a lot of moistening. To mix it in dry is to ask for trouble in the direction of uneven root action, so thorough moistening should first be effected by repeated sprinkling and turning with a clean shovel or spade on a clean concrete surface.

So much for the physical aspects of the ring compost; what of the plant food reserves?

Peat provides virtually no nutrients, clay scarcely more except, perhaps, in slow-acting reserve and the loam or garden soil will not be particularly rich in terms of ring culture requirements. It is very necessary, therefore, to add a base fertilizer and for this purpose that standard, universally known and available product, John Innes Base, is unexcelled. If the compost is being made up with the ideal heavy loam, thus avoiding the need for adding clay, J.I. Base supplies all the nutrients required in fertilizer quantities and in very suitable proportions for the balanced growth of tomatoes or, indeed, for virtually any other crop which might be grown by the ring culture method. When one adds clay the position is, I find, a little different because all the types of clay I have used possess a strong power of 'locking-up' phosphates in a way which makes this nutrient difficult for the roots to take up. As a result, if one merely adds J.I. Base to a clay-supplemented compost there is a likelihood that the young plant will be short of phosphates and this, in the case of tomatoes, reduces the vigour of root action and hinders setting of the early trusses.

For this reason, when using such a compost, I find it very desirable to add superphosphate to supplement the amount already present in the J.I. Base. My own procedure, which has proved quite successful over a number of years, is to add 12 oz. of J.I. Base and 4 oz. of superphosphate to each bushel of the soil/clay/peat mixture. It is most important, of course, to work these fertilizers evenly into the soil mixture. The best way to do this is to have the latter in a shallow, flat-topped heap on a clean concrete surface, to sprinkle the fertilizers evenly and then to shovel the heap backwards and forwards at least three times. This job just cannot be done too thoroughly.

The Ring Compost

What about lime? It will be remembered that the John Innes formula calls for the addition of a small amount of carbonate of lime—$\frac{3}{4}$ oz. per bushel for each 4-oz. 'dose' of J.I. Base—to compensate for the acidifying effects of the base.

If one is using a heavy loam, and therefore not employing added clay, and that loam gives a weak lime reaction in the test described below, it is certainly desirable to adopt the regular procedure. This involves adding $2\frac{1}{4}$ oz. of carbonate of lime per bushel to compensate for the 'three-dose', i.e. 12 oz., addition of J.I. Base.

When, however, clay is added to a light loam the picture is again rather different. I have never added lime under these circumstances for the simple reason that all the types of clay I have ever used, whether blue, grey or warm brown in colour, have proved on test to be distinctly alkaline, i.e. amply supplied with lime. If a lime-deficient clay were being used, the usual lime addition should certainly be made.

The home test for lime is very simple and well worth carrying out so that one knows for certain whether or not the standard addition is required. Put a teaspoonful of the soil or granulated clay into a wine glass and pour on just sufficient vinegar or old accumulator acid to wet the mass. Visible frothing indicates plenty of lime; slight bubbling or no reaction at all suggests that the lime addition is required.

Here, then, we have the complete ring culture compost: compensated if necessary for unduly light texture or loam or soil, fully supplied with the necessary nutri-

55

ents in the right proportions and with lime added if required.

All this may seem a lot of bother and, perhaps, unduly fussy but I can assure anyone with this thought in mind that the qualities of the ring compost constitute the most important single feature of growing by ring culture.

Most of the failures reported with this system can be traced back to an inadequately prepared compost, so do not be misled by those casual types who assert that they shovel any old garden soil into the rings and produce 20 lb. of fine tomatoes per plant. They might do—once—but the risk is too great and, after all, really quite unnecessary!

5

Early-Season Culture

------- >>►◄◄◄ -------

The aggregate has been laid down, the rings made or bought and the ring compost prepared. The time has come to plant.

Suitable spacing of the rings is shown in the diagrams. The natural temptation to plant more closely should be resisted as, far from producing a larger crop, a greater number causes each plant to be crowded by its neighbours with adverse effects upon the strength of the trusses and the distance between them. This applies no less to ring culture than to any other method of growing tomatoes; the best plants are always adjacent to the door where they receive a double share of air and light. I have fallen into this closer-planting trap myself more than once, hence the warning!

Set out the rings on the levelled surface of the aggregate and shovel in compost until each is roughly half full. Then, with a thick glass pint-size bottle or, more safely, with a similarly shaped piece of wood, ram the compost lightly. Turn the plants out of the 3-in. or 5-in. pots in which they have been grown to the truss-visible stage, support the ball of roots on the rammed surface and fill in further compost. Ram lightly, paying particular attention to the circumference of the ring and finish up with the compost half an inch or so higher up the stem than it stood in the pot. Avoid, of course, ramming above the ball of roots; the fingers must suffice in this instance. I usually snap off the seed leaves and lower small leaves at the turning-out stage to facilitate the job of planting in the rings.

The compost having been used in 'potting' condition of moisture and the root ball thoroughly watered the day before transfer, there now remains nothing but to give about 1 pt. of water per ring to complete the settling of the roots and the compost, and to provide a reserve.

I have laboured this job of planting somewhat because slipshod handling at this stage is, I find, one of the most frequent causes of over-dry compost later in the season; moderately firm ramming on moist—not wet— compost is of the utmost importance. The final level of the compost in the rings is not of great significance at this stage providing at least $1\frac{1}{2}$ in. of free space is left for feeding in due course. With plants turned out of 3-in. pots rather than the ideal 5-in. size, a little more compost will need to be put into the rings in the first instance to achieve a sufficiently high final level.

The purpose now is to encourage the roots to spread steadily out from the pot ball throughout the ring compost and down into the aggregate. This they will do most strongly and rapidly if they are searching for water so, however great the temptation to be kind to the plants, *avoid applying any water in the rings until the growth darkens considerably in colour and assumes a somewhat blue-grey tinge.* In sunny spring weather, this occurs after about a week though it may be nearer a fortnight in a dull period.

This is the signal for 'ball-watering' in the rings to the extent of ½ pint or so to each applied close to the plant stem every other day for the next fortnight or so and for commencing the daily drenching of the aggregate. This 'little and often' watering in the rings returns the compost to normally moist condition and establishes the necessary capilliary uptake from the continually water-charged aggregate.

With the roots entering the aggregate layer (easily visible around the bottom edge of the ring when the aggregate material is drawn gently aside), ball-watering should be reduced as the plants will be drawing most of their water requirements from beneath.

Once the first fruits are visibly swelling, ball-watering is discontinued completely and liquid feeding is commenced in accordance with the routine described in Chapter 6.

Now, the vitally important point to note about this immediately post-planting period is that the plants must not be treated too generously for water in the rings. Many newcomers to this way of growing have complained to me that their plants never made worthwhile roots in the aggregate. This can, of course, be due to an over-dry or toxic condition of the aggregate but more often it results from nothing more or less than keeping the compost roots so well supplied with water at the early stages that they have no need to extend down into the aggregate to find it.

Early-Season Culture

The customary daily light overhead damping of the plants once they come into flower does, of course, help to prevent the compost drying out unduly as some water always drips down through the leaves and the thorough firming at the time of 'potting' does the rest. Likewise this overhead damping moistens at least the surface of the aggregate and so encourages the roots to extend down from the rings.

To my mind, one of the greatest advantages of ring culture is the foolproof nature of the watering technique. When tomatoes are being grown in large pots, watering is quite critical because overwatering will bring about water-logging with consequent death of the roots, while underwatering results in the compost drying out suddenly and the plants wilting disastrously without warning. The same sort of thing can happen when the crop is being grown in the actual soil of the house and, in fact, there is no doubt that more plants are lost, or at least fail partially, from deficient watering than from any other single cause. In ring culture, these troubles do not arise since the sharply drained aggregate cannot become waterlogged and so can be watered to any degree of excess with nothing worse than waste of water.

Reverting to the subject of planting in the rings, it may be of value to some readers to record a modification which I have found most useful on occasion. It will be noted that the rings are filled *in situ*, i.e. after being placed squarely in the positions they are to occupy for cropping. In most cases, this will be the obvious method to follow and with a bottomless container, such as a ring, it will be apparent that it cannot be handled like a pot on the potting bench and then placed down on to the aggregate.

Nevertheless, one sometimes has boxes of pricked-out plants on part of the temporary propagating benches at the time when the tomatoes should go into the rings. They would be too heavily shaded beneath the benches so the only thing to do is to accommodate them for the time being on the bench. The method then is to plant as previously described but with the rings standing on squares of asbestos placed on the bench. When, in the fullness of the spring, the bedding plants can be put out to harden off and the benches removed, the rings can be slid off the asbestos on to the aggregate without disturbance of the compost or harm to the plants even though the first truss may then be in full flower. Placed ring-thick, all the plants for the house can be accommodated on a relatively small section of bench and, in fact, this elevated position helps early development of the growth as well as facilitating ball watering.

6

Feeding

————————— >>➤➲◐◖◄◄— —————————

Once the first-formed fruits on the first truss of the majority of the plants have swelled to about the size of a large walnut, routine feeding should be commenced.

Now, the purpose of feeding is to meet the nutritional needs of the rapidly developing plant by supplementing the plant food materials added to the compost at the time of planting.

At first sight, there would seem to be no reason why all the food required by the plant throughout the season should not be supplied at the beginning. This would be ideal as nothing would then be needed but to water pot plants regularly and to water occasionally in the ring culture rings to supplement the water sucked up from the continually moist aggregate. Unfortunately, this cannot be done as the quantity of base fertilizer required to be put into the compost for each ring would be so great that the roots of the young plant would shrivel up and die. For this reason, we must go about the job of supplying the food requirements by slow and steady degrees.

Feeding is the growing season counterpart of putting

base fertilizer in the compost and in ring culture we find it best to feed in solution rather than to apply a solid tomato top-dressing. The reason for this preference is simply explained. In ordinary pot culture, water is applied frequently to the compost and fertilizer spread over the surface is slowly dissolved and made available to the roots. As already stated, the watering technique in ring culture is different and there would be little opportunity for a solid feed to be carried down to the roots. Quite apart from this circumstance, solution feeding, i.e. with all the nutrients dissolved in water, is much more efficient as all the roots receive the same strength and balance of nutrients.

Ring culture 'grew up' at Tilgate on a proprietary liquid fertilizer concentrate known simply as '667' because it had an analysis of 6 per cent nitrogen, 6 per cent phosphoric acid and 7 per cent potash. This fertilizer also contained magnesium which made it particularly valuable for use on plants grown in limited volumes of compost where this essential food element is liable to become in short supply.

All manner of variations on this analysis were tried out and none gave better results than 667; indeed, it is difficult to conceive that anything could promote heavier cropping than the Tilgate 'demonstration' rate of more than 20 lb. of near-perfect fruit per plant in the course of an 8-month growing season.

Earlier work at Tilgate had shown that the ideal amount of 667 liquid fertilizer to give each plant is $\frac{1}{2}$ fl. oz. every 7 days throughout the main cropping period of around 16 weeks in the greenhouse or 8–10 weeks out of doors. With soil-grown plants, i.e. those set out in the border, this amount of fertilizer can be diluted with any desired quantity of water; in fact, the weaker the mixture within reason the better because the use of plenty of water ensures that all the widely extending roots are reached.

In ring culture, however, the feeding roots are concentrated entirely in the relatively small volume of ring compost so quite obviously the usual procedure of applying a big volume of weak solution has to be modified. After all, the one-fifth bushel or so of compost in each ring has only a limited capacity for retaining water or fertilizer solution. Careful experiments have shown this quantity to be a little under 3 pts. with a compost of ideal texture and which is in just-moist condition.

It seemed, then, that the $\frac{1}{2}$ fl. oz. of liquid fertilizer would need to be contained in about $2\frac{1}{2}$ pts. or water if all the fertilizer were to be retained in the compost. This gives a solution strength of $\frac{1}{2}$ fl. oz. of fertilizer in 50 fl. oz. of water, or in other words, a concentration of almost exactly 1 in 100. Really a very high strength, this, and one which, it was feared at first, might cause scorching of the feeding roots.

Careful trials with young tomato plants in 5-in. clay pots standing on a bench showed, however, that this strength of feeding solution did not, in fact, cause root scorching even when applied every time that water was required. It was apparent, nevertheless, that the plants were becoming much harder in growth and darker in the foliage than others alongside watered and fed in the normal manner for pot plants.

Feeding

With these observations in mind, ring culture plants were fed according to this high-strength routine. In this instance there was a much less marked effect upon the quality of growth; the balance of the plants was excellent and the trusses were sturdy and set well but there was no excessive hardening. Further investigation of this interesting circumstance showed that the absence of undue hardening was due to the fact that the plants could draw freely on the aggregate for their water requirements. By comparison, the plants in the small pots were having their ability to take up water restricted by the strong fertilizer surrounding all their roots. In other words, the double root system produced in ring culture very conveniently allows the optimum $\frac{1}{2}$ fl. oz. of liquid fertilizer concentrate per week to be applied in a quantity of water which can be absorbed by the compost.

To demonstrate the latitude in strength of feeding solution which ring culture permits, a batch of plants were, on one occasion, each given the full $2\frac{1}{2}$ pts. per week for several weeks of double strength mixture, i.e. 1 in 50. Even then, no harm resulted though, in the absence of any apparent benefit, it seemed that the additional fertilizer was being wasted.

Even today, despite all this careful work at Tilgate years ago and all the accumulated experience of the intervening years, one comes across recommendations for use of a particular liquid fertilizer in ring culture at a strength which is much too low for anything but plants growing in the wide expanse of border soil. As a result, the plants are half starved all through the season, the crop is light and the quality of fruit poor and, not unnaturally, ring culture as a way of growing is condemned as useless.

In my experience, more ring culture crops prove disappointing due to under feeding than from applying too much; there is no point in experimenting when everything has long since been worked out to the last detail!

Although in ring culture we feed at a comparatively high strength of fertilizer solution and the plants benefit thereby in terms of weight and quality of crop, it is important not to allow too great an accumulation of soluble materials in the ring compost. There comes a time, usually after 6 weeks or so of feeding, when the growth rate of the plants slows down quite noticeably and there may also be a tendency for some of the blooms on the higher trusses to 'cut', i.e. for the blooms to fall off even before they open. While a primary cause of this disturbing effect is a deficient supply of water due to poor rooting in the aggregate, it can also be caused by the joint effect of an accumulation of soluble residues from the fertilizer and a sharp increase in the acidity of the compost due to gradual destruction of lime by the acid-forming components of the fertilizer.

Some years ago, I investigated these circumstances in some detail by means of soil analysis and I was able to work out a simple technique for avoiding the potentially harmful effects of continued heavy feeding. In brief, I found that a level tablespoonful of carbonate of lime applied to each ring two or three times in the course of the feeding season—say, after every sixth feed—effectively overcame the tendency for the pH of the compost to fall to an undesirable degree. Now, this lime has to be carried into the compost and this needs a greater quantity of fluid than is given by the weekly feeds.

I therefore worked along the lines of filling up the free space of the rings several times with plain water after applying the lime. Mid-week between two feeds was the obvious time to do this and one could go round filling up the rings with water and, by the time the last rings had been dealt with, the first were drained and ready for the next dose. This heavy and repeated watering over a short space of time also had the effect of flushing the soluble residues from the compost into the aggregate where, in turn, they would be dispersed by subsequent drenching. The result, then, was to carry in some at least of the lime and to flush the compost.

I found the effect upon the growth of the plants and the quality of the fruit so good that I have adopted this procedure ever since and I strongly recommend it. Quite a lot of the lime, by the way, is left on the surface, and shows up white when the top of the compost dries out but, bit by bit as the subsequent feeds are applied, this residue is carried down.

To recap, then; in ring culture feeding is commenced when the food reserve provided by the base fertilizer is diminishing but before weakening of the top growth indicates a definite shortage of nutrients. With plants set out in gentle heat towards the end of March, my own crop usually reaches the stage for feeding 4 weeks after planting, i.e. by the fifth week-end since, as with most amateur gardeners, the major jobs get done mainly at week-ends.

Although in practice there seems to be no objection

to going right over to the full rate of feeding, i.e. $2\frac{1}{2}$ pts. of solution per ring once a week as soon as the ball-watering phase is completed, I prefer to start cautiously by giving a half-quantity feed over the week-end and the remaining $1\frac{1}{4}$ or so pts. per ring mid-week. Thereafter they have the full dose and I feel instinctively that this is the right way to go about the job.

Some ring culture growers carry on right through the season with half-quantity twice-weekly feeds as they find that this keeps the compost moist more effectively than the longer spaced full feeds. I have, in fact, suggested this procedure many times where there have been reports of the plants suffering through an over-dry compost, but it is really necessary only where the compost is too loose and sharply drained and so is unable to absorb the full dose at one time. I suggest that readers try out the two methods side by side, perhaps half the crop on each, and see for themselves whether there is anything to be gained by the lighter, more frequent feeding under their own conditions.

Most gardeners are happy to use proprietary feeds because, although they do involve quite an expenditure in the course of a season, no 'messing around' is required and one can be sure of an unvarying analysis and uniform results. I have referred to 667 liquid fertilizer because this is the feed on which ring culture 'grew up' and the one which I personally have found unequalled over the years. Unfortunately this fertilizer is not as widely available through retail sources as it deserves to be and the same applies to the more recently introduced 'Triple 667', the soluble solid form.

Feeding

However, where supplies can be obtained, the feeding mixing routine is delightfully simple. With 667 liquid fertilizer, 3 fl. oz. (6 medicinal tablespoonsful) are stirred into a 2-gal. canful of water. This, in fact, gives a feeding strength of a little less than 1 in 100 but the quantities are convenient and the difference negligible. In the case of Triple 667, since, as the name suggests, the concentration is three times as great, the equivalent dilution rate is 1 oz. in 2 gal. of water. The solution, produced by thorough stirring, is, by the way, a fine strong green colour whereas the liquid fertilizer colours the water no more than a pale brown. An unimportant point, this, but perhaps worth recording for the sake of the newcomer who may wonder whether all is in order when a reddish-brown solid dissolves to give a green solution!

Compure 'K', Boots' tomato liquid fertilizer, has a somewhat similar analysis and is used by many gardeners successfully for ring culture feeding; but it does need to be used at the regular ring culture strength rather than at the lower rate advised for normal border or pot growing.

Proprietary feeds are convenient but, as I have said, rather expensive to use in quantity. This accounts for my having been asked on many occasions whether ring culture feeds can be made up at home from easily obtainable fertilizer ingredients.

A few years ago, I carried out a series of trials to see what could be done in this direction and the final outcome was a feed which gave results consistently indistinguishable from 667. It certainly involves the bother of obtaining and mixing the several fertilizer ingredients, it needs more stirring into water to dissolve the soluble part of some of these ingredients and, as it produces a more or less colourless solution, there is a deceptive impression of weakness, but it certainly does the job at about one-quarter of the cost of the proprietary types!

The procedure is to mix together:

> 4 parts by weight nitrate of potash
> 10 ,, ,, ,, superphosphate
> 5 ,, ,, ,, sulphate of ammonia
> 7 ,, ,, ,, magnesium sulphate

The unit of weight can, of course, be ounce or pound. The mixed fertilizer has an analysis of almost exactly 6 per cent nitrogen, 6 per cent phosphoric acid and 7 per cent potash together with a good supply of magnesium so, since the actual soluble analysis in terms of nitrogen, phosphates and potash is the only thing that counts where a soluble concentrate is concerned, it is understandable that such a mixture will give results comparable with the manufactured 667 feed.

To use this feed, dissolve $1\frac{1}{2}$ oz. in each gallon of water with the help of very thorough stirring. The solution remains cloudy due to the presence of the only very slightly soluble gypsum in the superphosphate but this need cause no concern as it does not affect the strength of the feed. The solution produced is employed in exactly the same manner as a diluted proprietary feed i.e. at the rate of $2\frac{1}{2}$ pts. per ring every 7 days or 1 pt. every 3 days.

The following points should be noted when making

up this feed: make sure that nitrate of potash or agri-
cultural saltpetre is obtained (analysis around 13·5 per
cent nitrogen, 45 per cent potash) and not the totally
different and, for this purpose, unsuitable potash nitrate
or Chilean saltpetre.

Crush separately each ingredient very thoroughly and,
if a hair sieve is available, remove all lumps for further
crushing. Mix the ingredients by stirring in a dry bucket
or, for small quantities, by repeated turning on a sheet
of paper. Keep the mixture dry by putting it into a polythene bag which, in turn, is contained in a tin. Label the
tin, 'Ring culture feed—1½ oz. per gallon', to avoid any
doubt later on.

Since I gave details of this feed in my ring culture
series in *Popular Gardening* several seasons ago, many
readers, according to their end-of-season reports, have
used it with complete satisfaction. However, every true
gardener is cautious about going over to something
which he has not tried first so I suggest that this home-
made feed be used, parallel to a proprietary type, on
part of the crop only for the first season.

Another useful tip on feeding applies mainly to the
early season. Some gardeners, with their own very
definite ideas on making up soil composts, are much too
generous with nitrogenous materials such as bone meal
or hoof and horn meal. As a result, by the end of the
ball-watering period, the plants are inclined to be coarse
and leafy and have much more emphasis on growth than
on fruit.

Such plants need the balancing effect of potash on its
own and, in ordinary pot work, where water application

to the compost is continued, this can be achieved by sprinkling a level teaspoonful of sulphate of potash over each pot and leaving the subsequent waterings to carry it down to the roots throughout the pot. It will be obvious that this procedure does not fit in with ring culture so we make up a sulphate of potash liquid feed.

This type of potash is neither highly nor readily soluble though a solution can be produced containing around 1 lb. of sulphate of potash per gallon of water at normal springtime air temperatures. Something between 1 and $1\frac{1}{4}$ lb. of the ordinary fertilizer grade is put into a gallon of water in a bucket or can and the mixture is stirred whenever one happens to be around during the next day or two. All the soluble portion dissolves leaving the mineral impurities as a deposit. A suitable rate of use of this solution is $\frac{1}{4}$ pt. per gallon of water and each ring should be given about 1 pint of the diluted purely potash feed.

Usually only one or, at the most, two such feeds are needed to change a soft, pale, sappy plant into one with the desirable firm, bluish-grey foliage and strong truss development. Once the plants have 'steadied up', normal ring feeding can be adopted. The sulphate of potash solution keeps indefinitely so any surplus can be stored in a bottle (duly labelled!) for use on future occasions when ring culture tomatoes or, indeed, any other plants, show the typical signs of undue softness.

Here, then, is the established routine of ring culture tomato feeding.

If these notes seem unduly single-minded and emphatic, my reply is that they are intended to be so. There

Feeding

can be little purpose in adopting a concisely planned system of growing if feeding, one of the main factors determining the success or failure of the crop, is treated in a casual manner.

7

Ring Culture Tomatoes in the Garden

————————▸▸▷●◁◂◂————————

Most gardeners seem to look upon outdoor tomato growing as a doubtful proposition and, having regard to our fickle climate, this is, perhaps, scarcely surprising.

From my own experience in the home counties, it is only in an exceptionally good summer, such as that of 1959, that one can expect to ripen a worthwhile crop of fruit from plants set out across garden or allotment.

Results are, however, much more reliable when the plants are given the warmth and protection of a south-facing wall or fence. In view of the material degree of frost and wind protection afforded by the background, planting can be several weeks earlier than normally considered safe and this circumstance, together with the more rapid growth promoted by the radiated warmth, makes for cropping while the fruit is still expensive to buy. Tomatoes in July, especially home-grown, still have the appeal of novelty; in September, they can scarcely be given away!

By planting in such a situation in the border soil, the first fruits will, on average, ripen a good fortnight earlier than those on plants set out in the open. If culture is in pots and the plants are brought on for a while in the greenhouse in these pots, a further fortnight can be gained and ripe fruit in late June is by no means unusual given even a reasonable amount of sunshine.

At one time, before ring culture was evolved, I grew tomatoes in large clay pots against the background of a high, south-facing fence and did very well with them but keeping 'on top' of the watering was a considerable job in hot spells while early-season watering was a most critical business if disastrous root damage was to be avoided.

With this background experience—in more senses than one—I turned over to ring culture for my outdoor crop without hesitation because it was apparent that here was a means of achieving results at least as good as with the most careful pot culture and with watering reduced from a critical, day-in and day-out job to a simple, virtually foolproof routine. The results over the last ten seasons have more than justified my early optimism and I can unreservedly recommend outdoor tomato growing by ring culture to every gardener who has in existence a warm site or who is prepared to contrive such a site.

This, as I know from my considerable mailbag each season, applies scarcely less to northern England and even to Scotland than to southern gardens though, obviously, the warmer the climate the earlier will the first fruits ripen and the greater will be the crop of ripe fruit before the autumn frosts call a halt.

Grown as a single row of plants right to the top of the fence or wall, tomatoes are by no means out of keeping with normal flowering subjects in the foreground and this aspect of the crop is illustrated in Frontispiece. Certainly there is an element of novelty as well as of utility about such an arrangement and, if one sets out to obscure the unsightly blankness of a wall or fence, it might as well be with a decorative food-crop as with flowers pure and simple!

Outdoor ring growing involves exactly the same procedure as with the greenhouse crop. The aggregate layer, anywhere from 4 to 6 in. deep, can either be laid in a trench cut hard against the fence, and 15 in. or so wide, or it can be entirely above ground level with a retaining board or row of blocks. This latter arrangement is convenient where the foundations prevent a clean trench being dug and also where a path rather than a border runs adjacent to the fence or wall.

In the case of a trench aggregate, it is advisable to have the surface of the medium $\frac{1}{2}$ in. or so *above* the general border level to minimize the percolation of soil into the aggregate during the course of watering. An alternative is to separate the aggregate layer from the border soil by a board projecting an inch or so above the general level. This arrangement, though more bother, is to be preferred as it prevents the water displacing either aggregate or soil.

Since there will be only a single row of plants, a sufficient spacing is 15 in. from stem to stem, i.e. from ring-centre to ring-centre when the rings are positioned for filling. More remote spacing may result in slightly superior individual plants—those at the row ends tend to be stronger—but since even in ring culture one cannot expect to ripen more than five or six trusses and the early trusses are the most valuable, it pays to go for the maximum number of average-good plants rather than fewer superb ones.

It is impossible to give anything precise in the way of a safe date for planting since season, district and elevation of the site in relation to the surrounding land all come into the picture where spring frosts are concerned. In the London area, I habitually plant at the end of April and have never yet lost a plant from frost. The risk is obvious, however, and I make a point of keeping handy a 4-ft. wide roll of hessian which can be secured along one edge to the fence with heavy drawing pins and draped down over short vertical canes set in the ground in front of the plants. In this way, crushing of the foliage is prevented and several degrees at least of frost can be kept off.

An alternative, most useful in cold windy weather since the plants can be left covered for as long as may be needed, involves a light transportable wooden framework into which sheets of glass can be fitted to produce a 'half-barn' cloche type of structure (Diagram on p. 28).

Even without the provision of protection other than the wall or fence, it is certainly worthwhile to plant 3 weeks to a month earlier than one would dare to do across the open garden. The potential gain in earliness and overall crop of ripe fruit more than outweighs the risk of losing the plants during the first week or two.

In view of the possible necessity to cover the plants, it is as well to support them for a start with ties to short canes inserted in the rings at planting time. Canes of

30-in. overall length are suitable for this purpose. Subsequent support can be provided either by means of four-ply fillis string, as described in Chapter 8 for plants in the greenhouse, or by 6-ft. canes. This latter method is certainly the more desirable as wind-sway is prevented, but the cost involved is quite considerable. If it is intended to use these tall canes, it is important that the short canes inserted for early support be of such thickness that, when they are released from the plants and withdrawn, the holes left in the ring compost are large enough to allow the tall canes to be slipped in without tearing the roots. Top support for these canes is afforded by tying each to a wire stretched tightly between 9-in. brackets firmly secured to the fence or wall.

To avoid any possibility of heavily-laden plants slipping down the canes and kinking the lower stem, ties between plant and cane should be made every foot or so. Four-ply fillis is suitable for the purpose as it is strong yet non-chafing, and ties should be made first around the cane and then around the plant beneath a leaf stem.

I normally stop the plants one leaf above the sixth truss or if growth has been particularly good, above the seventh truss. In 1959, seven trusses ripened virtually to the last fruit; in 1960, five trusses ripened and one was picked mostly green in early October.

The watering procedure with the outdoor crop follows the same general lines as in the greenhouse but, whereas shading of the grass helps to reduce water demands of the indoor crop in heatwave periods, the open-air plants are subjected to the full effects of drying sun and wind. As a result, watering in the rings mid-week between feeds is inescapable if the plants are to be kept happy but this is a good deal less demanding than the once or even twice a day watering required by pot plants under similar conditions! One of the factors making for increased water demands out of doors is the direct playing of the sun on the rings. It is helpful in this regard to plant heavy-foliaged flowers, such as bedding dahlias, immediately in front of the rings to provide shade while, where the rings border a path, a row of old asbestos roofing tiles slanted against the rings provides a useful insulating effect. A small point this, perhaps, but it saved me a lot of water-can work in the otherwise ideal summer of 1959.

The garden crop is fed in accordance with the standard procedure but, as the plants are growing for a shorter period than in the greenhouse, 8 to 10 weekly feeds are usually sufficient to mature the crop.

Passing to more general matters, the resistance to disease of the outdoor ring crop is often a subject of gratified comment on the part of gardeners writing to me. Certainly my own crop very seldom shows any sign of the fungus troubles usually associated with outdoor tomatoes. Even in the persistently wet summer and autumn of 1960, that habitual bane, potato blight disease, appeared only as October advanced. By comparison, traditional outdoor tomato crops all around had been so riddled with this fruit and foliage disease that nothing worthwhile could be picked even in green condition.

Why should identical varieties grown by ring culture exhibit this marked immunity?

Ring Culture Tomatoes in the Garden

First and foremost, I think, is the fact that the general system of providing a steady water supply, coupled with regular high potash feeding, makes for the ideally balanced, firm yet vigorous growth. Then there are the circumstances of culture. At the critical disease period of late summer, the first three or four trusses of fruit have been picked and the lower foliage removed. In other words, the actively growing part of the plant is well up in the air and away from the rain-soaked ground. Then, of course, the wall or fence dries out rapidly after rain and generally keeps the air relatively dry round the fruit and foliage. Lastly, the effect of the solid background is to induce air currents and thus to prevent the stagnant conditions so favourable to the entry and establishment of disease organisms.

Birds, and particularly the ubiquitous house sparrow which seems to be concerned with nest building throughout the summer, have a mischievous habit of pulling away at odd ends of string. This is annoying enough with border plants tied to stakes and canes but it can be disastrous where the string method of support is used for the ring culture plants since the top looped ties may be frayed and torn to the point where the slip knot gives way and the laden plants collapse. A permanent knot at this top wire is not appropriate as every now and again the string must be disengaged from the wire, the plant being carefully supported meanwhile, and the 'stretch' taken up. The only thing to do is to wax the loose end with a tallow candle to make it less responsive to pecking!

After being stopped, outdoor plants, in common with

those under glass, tend to produce side shoots at an amazing rate from the area immediately below the point of stopping. Now, whereas throughout the main growing period all side shoots should be removed while still small by snapping them cleanly from the leaf axils, something less drastic is desirable after the leading growth has been topped to assist in maturing the trusses already well formed. The sap continues to rise strongly especially after feeding and, without the provision of a 'growth' outlet, the fruit has to bear the brunt and is often caused to crack by the sheer pressure of sap within. This tendency can be offset by stopping the side shoots themselves after three or four leaves have been formed and thus encouraging further vegetative growth, though at a slower and less demanding rate than that of the original head of the plant.

As soon as the leaves are touched by frost in the autumn—usually those on the shoots extending above the wall or fence are first affected—they should be stripped off and the trusses bearing near-mature but still green fruit hung up in a shed to ripen. There will be less shrivelling of these fruits if a couple of feet or so of the main stem is retained above and below the trusses. Alternatively, the green fruit can be used for chutney.

The rings, though usually still fairly sound, will not stand up to a further season so they should either be burnt and the ring compost stacked for the following year's final chrysanthemum potting or, with the tomato rootstock pulled out, planted up with tulip bulbs for spring flowering in the greenhouse.

8

General Culture of Tomatoes

Although, as will be apparent from the foregoing chapters, ring culture is fundamentally different in certain aspects from the normal ways of growing, there are many common features as regards general culture of the crop.

Our concern, then, is with cultural operations which are in no way particular to ring culture but which apply equally however the plants are grown.

PLANT QUALITY

First and foremost, set out the best possible plant which you can raise or purchase, as the case may be. A sturdy plant with a spread as great as the height, with the seed leaves still healthy, with dark green rather than pale foliage, with the first truss clearly visible and with a ball of thick, white roots, will always give a better account of itself than a spindly, hard, half-starved plant taken from an overcrowded box. Unless it is diseased, the poor quality plant will eventually 'get away' if the subsequent culture is good but, by the time it does, the good plant is setting fruit!

I like to grow my plants on in 5-in. pots for a fortnight or so rather than plant from the usual 3-in. size. There is certainly extra work involved but the plants develop more rapidly and crop appreciably earlier. As regards compost, J.I.P.2 is ideal both for the 3-in. and 5-in. pots.

If the plants are indeed grown on in 5-in. pots, they will need support before being set out and this can be given with thin canes about 15 in. long and a single tie of raffia or soft string. Where they are set out from 3-in. pots or from boxes, the canes should be inserted in the rings at that stage.

DAMPING

Overhead damping with a syringe or fine-rosed can should be given mid-morning on bright sunny days as this application of water to the foliage has several desirable effects. First and foremost, setting of the blooms into fruit is promoted. The effect here is that the water falling on the plants causes a gentle movement and thereby shakes the pollen from the ripe anthers on to the central stigma of the bloom. Furthermore, the evaporation of this water in the warm sunlight creates the humid atmosphere required for the pollen grains to grow and to effect fertilization. These delicate little structures rapidly lose their activity in dry air and this accounts for the poor set often experienced in sudden heatwave periods when extra ventilation is given to prevent the temperature from rising unduly and the air around the plant becomes much too dry.

Secondly, the moist air around the recently damped plants reduces transpiration of water by the leaves and so eases the strain on the roots at the hottest and most demanding period of the day.

Finally, moist air discourages that most difficult to control of pests, the red spider mite.

It is apparent, then, that there is much to be gained by at least one damping of the developing plants on sunny days.

As growth progresses and an appreciable part of the volume of the greenhouse becomes filled with leaves breathing out moisture, this overhead damping is no longer necessary and, indeed, should be discontinued in the interests of avoiding Botrytis rot of the fruit, leaves and stems.

SUPPORT

The use of tall canes for supporting the plants is described in Chapter 7 and the system can be employed for both outdoor and greenhouse crops.

Plants growing under glass are, however, protected from wind-sway so there is less need for the rigid support provided by canes. I would, personally, incur the cost of canes only if the greenhouse roof members were too weak to take the weight of the crop though, even in this case, a wire attached to the roof can be avoided by erecting angle-iron posts at each end of the greenhouse and running the top wire between these posts.

The least expensive, though nevertheless thoroughly

satisfactory, method of support is by means of a string running from the lower stem of the plant to an overhead wire. The soft but very strong four-ply fillis is suitable for this purpose and the procedure is as follows:

Measure the distance between compost surface and overhead wire, add on about 6 in. to be used up in making knots, and cut the required number of strings to this overall length. Secure one end of a string to the plant stem by means of a non-slip knot, allowing plenty of room for the stem to increase in girth as the plant grows. This lower point of attachment must be immediately below a strong leaf-stalk and I usually make the tie a couple of leaves or so beneath the first truss since a leaf at this point will be retained well on towards the completion of cropping whereas the lowermost leaves are removed once the plant is established, as they interfere with watering. The other end of the string is then taken up to the wire at a point vertically above the plant, two turns are made round the wire and a slip knot tied. The weight of the plant holds this knot but it can be undone quickly when the necessity arises from time to time to take up stretch in the string.

Be careful not to have the string more than just taut at this stage as any shrinkage on a really tight string can cause the plant virtually to be pulled out by the roots!

With the bottom end of the string secured to the plant and the top to the wire, the stem of the plant is entwined with the string by twisting the plant round the string. The best time to do this job is during the hottest part of a sunny day as the leaves and leaf stems are then pliable and resistant to being snapped off. Immediately prior to this 'twisting' operation, the short cane which provided early support should be removed.

Thereafter the entwining of stem and string is carried out every 5–7 days so that straight, vertical growth may be achieved. Here again the job should be done when the growth is pliable or the whole head of the plant may be snapped off.

An alternative to making the bottom tie to the plant itself is to provide a tightly stretched horizontal wire immediately above the rings and to secure the strings to this wire. In my experience, however, this method has no advantage and is an added complication.

SETTING THE BLOOMS

A tomato plant grown according to the correct ring culture procedure, as detailed in the foregoing chapters, is vigorous yet balanced in growth and this makes for even setting of the blooms. Nevertheless, if the temperature level in general, and the night temperature in particular, is much below 55 degrees for more than a few hours, the blooms then at the point of setting may fail to be fertilized by the pollen and a 'dry-set' results. The effect then is for the bloom to fade and drop off normally, for the calyx to remain green and fresh but for there to be no sign of swelling of the little pin-point fruit. It is unimportant if this happens with only the occasional bloom on a truss but when it is apparent that a high proportion of the blooms are not setting, something must be done if the weight of crop is not to be adversely affected.

General Culture of Tomatoes

Any obvious cultural faults, such as an over-dry condition of the ring compost, a dry atmosphere or an unduly low night temperature should be put right and, in addition, the trusses showing signs of faulty setting by the time the third or fourth blooms have dropped should be treated with a 'hormone' setting solution.

These hormone solutions, freely available at garden shops under such names as 'Fulset' and 'Tomato Set', contain harmless organic preparations closely allied to the natural plant hormones which are produced when fertilization by the pollen grains occur and which cause the ovaries of the bloom to commence swelling into the familiar tomato fruit. When applied to a bloom which has failed to set naturally, they stimulate the growth of a fruit which is normal in every aspect except that it is seedless.

Since in an amateur greenhouse one tends, with an eye to the cost of heating, to keep the night temperature in the early year much too low, and so invites a deficient natural set, I make a practice of treating all the trusses on each plant with a hormone preparation to ensure a good 'set' rather than to correct a bad one.

These preparations require to be applied only to the actual trusses and to be kept off the rest of the plant as far as possible since they stimulate distortions of growth, especially of the young leaves around the growing point. For this reason, a scent spray is commonly advised for applying the solution to the trusses but I find it much easier and a lot more certain in effect to have the solution in a small basin and to bend each truss gently over into the liquid.

Treatment is best carried out when the truss is in full flower and, providing the concentrated preparation, as bought, is diluted strictly according to instructions, the result is, in my experience, entirely satisfactory. Admittedly, making up perhaps half a pint of mixture to treat a dozen trusses is wasteful of material, especially as it must be used fresh and cannot be kept in diluted form but, even so, the cost is slight compared with the value of the extra fruit ensured by the treatment. Fruits able to set naturally will do so whether hormone treated or not, so the treatment is entirely advantageous.

Incidentally, avoid the temptation to employ a strength of solution greater than that advised on the label, since over-dosing is liable to cause 'boxy', i.e. hollow, fruits.

Tomatoes without seeds may sound unnatural but they are in fact no different in flavour or texture from those with seeds and they are certainly a great advantage for folk with artificial dentures!

To summarize: setting of the blooms is aided by gentle agitation of the plants—by overhead application of water or by tapping the wire to which the strings are attached—to release the pollen, and by creating a moist atmosphere during periods of bright sunlight when growth is most active. Natural setting may be replaced or supplemented by 'hormone' application.

REMOVAL OF LEAVES

All the while a leaf remains green and healthy, it is producing food for the plant and should therefore, in general, be retained. However, as each truss matures and begins to ripen, it will usually be found that the leaves immediately below are taking on a somewhat yellowish tinge and have obviously completed their useful life. Clean removal of the whole leaf by cutting hard against the main stem with a sharp knife is then desirable. An alternative method of removal is to snap the leaf away cleanly by bending it sharply downward, pressure being applied close to the point of juncture with the main stem.

Severe leaf stripping, extending over several partly mature trusses, though commonly practised commercially to hasten ripening, is undesirable since fruit size is adversely affected and, due to exposure of the fruit to direct sunlight, the hard 'greenback' condition may be promoted.

Once a truss is picked completely, the leaves may be stripped off up to the next truss to promote improved air circulation and to reduce the risk of attack by Leaf Mould and Botrytis diseases.

Training the Top Growth

Plants grown in pots or rings on raised benches rather than on the floor of the greenhouse require to be trained up parallel to the roof slope to compensate for the lack of height to the gable wires. This can be done by running further strings from these wires to a batten secured by brackets immediately below the ridge of the house. The growing point of each plant may be stopped as it approaches the ridge or, by the use of further supporting strings, induced to grow down the opposite roof slope, thus producing a complete arch. To avoid a dense tangle of growth, all side shoots must be removed while still small and also complete leaves here and there. With early-planted, floor-level crops, the same procedure can be followed to advantage in houses of limited height.

VARIETIES

Tomato varieties are legion and, while each has its advocates, none is ideal in all characteristics and under all circumstances of culture.

I have myself tried out a fairly wide selection of both the standard market and the purely amateur varieties but none has given me justification for going away from that fine old variety Stonor's Moneymaker. In most respects, it is the ideal gardener's variety as it sets evenly to produce substantial trusses of medium size, solid fruits which ripen evenly from 'white-green' to a clear, bright red. Well grown fruit has an attractive, mildly acid flavour and the central stalk 'core', so prominent with some varieties, is small to the point of insignificance. In my experience, this variety has only one failing: the fruit matures rather slowly and it does not come into bearing as early as some.

I have grown Moneymaker in the open ground, in clay pots, in the greenhouse border, by ring culture both outdoors and under glass. It maintains my high regard whatever the method of growing and is certainly at its best in ring culture. This, in fact, is a general finding as ring culture seems to suit all varieties though by no means all are worth growing even by this method.

For those who prefer a more highly flavoured tomato, I recommend Ailsa Craig. Though usually looked upon as being essentially a heated glasshouse variety, I have seen excellent outdoor crops of 'Craig' in hot summers and it is worth growing a few plants in the garden ring culture installation together with the main batch of Moneymaker. It is an extremely vigorous variety with heavy foliage, of moderate cropping capacity and with a tendency for wide spacing between the trusses. By reason of these characteristics, it is inclined to be somewhat overpowering in the small greenhouse and I would advise supplementary sulphate of potash feeding in the early season to keep the growth on the hard side and to restrict the spread of the leaves. Even then it is desirable to increase the spacing between the plants from 18 to 24 in. To my mind, well grown Moneymaker leaves nothing to be desired but I have to admit that Craig wins where flavour is the first and foremost consideration.

GREENHOUSE INSULATION

The free availability of recent years of thin, almost glass-clear polythene sheeting at modest price has made greenhouse insulation a thoroughly practical proposition. Both for propagation as a whole and early season culture of tomatoes, lining a house with this sheeting provides the advantages of a moist 'growing' atmosphere and considerable saving in fuel costs. Thus, I find that the watering required by seed trays of plants in general is reduced by half and that a half-kilowatt panel heater is quite sufficient to maintain a growing temperature in an 8 ft. by 6 ft. greenhouse when fully lined. The increased temperature lift is at least 10 degrees as between a lined and unlined house.

As spring advances into summer, the lining is no longer required and is, indeed, a disadvantage with tomatoes in view of the high humidity induced. (See Chapter 10, 'Botrytis'.) For this reason, I make a practice of removing the lining towards the end of May since, by such time, heating can be discontinued.

Lining is a simple operation as the sheeting can be stapled to the glazing bars of wooden houses or draped over wires stretched near the glass with the metal-framed type. The sheeting must be applied sufficiently taut to prevent sagging on to the glass as it is the trapped air between sheeting and glass which provides the insulation, not the sheeting itself.

The initial cost of lining is in the order of one new pence per square foot of internal surface covered but, by careful application and removal, the sheeting can be used for several seasons.

9

Other Plants by Ring Culture

----▶▶➤0⊂◀◀----

Readers turning to this chapter will expect to find details about growing a range of plants by ring culture. They will, I fear, be disappointed since in all sincerity I cannot specifically advise this system for anything but tomatoes.

Inevitably, when something new comes along, especially a novel and advantageous way of growing a crop, there are attempts by one and all to widen the scope; to open up exciting new prospects for better and easier growing of everything in general. This happened with tomato ring culture within a season or two of my announcing to the gardening world the findings at Tilgate.

CHRYSANTHEMUMS

Understandably, perhaps, the first plant to receive attention by the 'extensionists' was the chrysanthemum. Here was another subject habitually grown in pots by the gardener, so, why not chrysanthemums by ring culture?

In due course, blooms were shown in a television

gardening programme, the claim being made that the plants were taller and the blooms earlier than under normal pot culture. These results might sound impressive until it is remembered that, with most varieties of mid-seasons and lates, we have already to battle *against* inconveniently tall growth, and that we do not want the blooms any earlier than their normal season.

This corresponds with my own experience that no advantage can be claimed for ring culture at least as far as performance of the plant is concerned. No very serious criticism here, perhaps, for anyone who has a mind to grow the crop by this method, but now we come to an aspect of the project which renders it laughable to any who have a serious interest in growing chrysanthemums for flowering under glass.

Throughout the summer, the plants are grown in rows out of doors and, to make the late September job of housing easier and less of a check to growth, we habitually restrict rooting into the path beneath the pots by standing the latter on tiles, boards or polythene sheeting. According, however, to the ring culture protagonists, the plants are set in rings standing on the usual sort of aggregate layer and are fed and watered much as for tomatoes. Then, at housing time, the bottom roots are 'disengaged' from the aggregate and the whole issue, growth, ring and roots dangling beneath, is taken into the glasshouse and the aggregate roots settled back into a shallow trench cut in the greenhouse aggregate.

Even supposing it were possible to retain a worthwhile proportion of the aggregate roots—and, believe me, it is an awkward, unsatisfactory job—do we want the plants to have a big water-absorbing root system after they are housed? Most emphatically, we do not!

The point here is that, apart from the employment of an uneconomic amount of heat in conjunction with ever-open vents, the only way to prevent loss from damping of the blooms of most varieties in stagnant weather is to dry out the roots to the point of incipient flagging of the growth. This certainly cannot be done if there is a big reserve of water beneath the compost, so, here again, the project falls down.

The problem of retaining the outdoor aggregate roots at housing time could, of course, be overcome by having the rings standing on boxes of aggregate and transferring the whole issue but the weight involved would demand the assistance of an army of willing friends.

Certainly the job of skilled watering would be much reduced with rings as compared to pots since frequent soaking of the aggregate could be left to even the least horticulturally-minded member of the household; but this feature scarcely compensates for the obvious snags of the project.

No, most definitely, ring culture does not 'fit' chrysanthemums!

Other crops suggested as grown to advantage by ring culture are the cucumber and the perpetual flowering carnation. I have made comparative trials as between ring culture and traditional methods with both plants and offer the following considered comments:

Other Plants by Ring Culture

CUCUMBERS

Here we have a crop known to require an abundant supply of water at all times and, as such, one which should be suited to ring culture. In the first trial, the plants were treated precisely as for tomatoes in respect of type of ring compost, watering and feeding. The result was not encouraging: the growth was abnormally hard, the leaves were small, slow to develop, subnormal in size, and the fruit was inclined to have a bitter, harsh flavour.

In the light of this experience, the next trial involved the use of a much 'softer' feed, i.e. one high in nitrogen and phosphates and low in potash. In a parallel trial, feeding throughout the season was exclusively nitrogenous, involving the convenient organic source, soluble dried blood. Growth in both trials was certainlx more of the type normally associated with this luxuriant 'tropical' crop but the weight and quality of the fruit was still well down on traditional culture in which the plants are set in a bed of stable manure and turfy loam.

Later trials confirmed these findings so, while there is no doubt that the cucumber *can* be grown reasonably well by ring culture providing the balance of feed is radically changed, there seems to be nothing to gain. The results, indeed, do not come up to those quite easily and consistently achieved by normal manure-bed culture.

CARNATIONS

The perpetual-flowering, 'p.f.', carnation is quite

commonly grown in pots over a period of 18 months or more and thus might conceivably respond to ring culture. My own trials with this plant have been distinctly limited, mainly because the earlier results were not such as to encourage further work. Over a period of 12 months, the number of blooms produced per plant was no greater than by pot culture, the flower quality was not visibly improved and, on the debit side, the growth was softer and taller and less easily held within the limits of a small greenhouse. This finding was not unexpected as the carnation is naturally a slow-growing plant and one which has a small water requirement compared with the tomato. Here again, ring culture can be employed but not, it appears, to advantage.

LETTUCE

In the earlier days of the Tilgate work, quite good lettuces were grown in small paper 'collars' filled with J.I.P.2 compost and standing on an ash aggregate previously laid down for tomatoes. As might be imagined, however, considerable difficulty was found in feeding the plants once the leaves spread beyond the confines of the collar and this project was recorded as being of no more than academic interest.

More recently, I have had excellent results by what could be considered a modification of ring culture. The seedlings are pricked out into large soil blocks—about 3 in. in height and diameter—made up with J.I.P.2 compost. After 3 weeks or so on the greenhouse bench to allow the plants to establish, the blocks are buried to about two-thirds of their depth in the aggregate, 8 in. apart in a row in front of the tomato rings.

These large soil blocks contain all the food required by the plant to grow to maturity and the frequent drenching of the aggregate keeps the blocks sufficiently moist to overcome the need for individual watering. With the lower leaves clear of the moist aggregate surface, I find that losses from Botrytis are virtually non-existent and I am much impressed with the size and quality of lettuce grown in this way. With a suitable choice of variety, a winter crop could be grown to follow normally housed chrysanthemums and to precede the ring crop of tomatoes.

This covers the range of crops, grown by the gardener in his greenhouse, which might seem even remotely suited to ring culture. None, apart from the instance of lettuce, subjected to a modification of the system, appears to respond really satisfactorily and I think it is much more realistic to look upon ring culture as essentially a tomato-growing system.

It excels as such; why try to widen the scope to encompass less demanding crops which can be grown better by traditional methods?

10

Common Troubles

I do not propose in this specifically ring culture book to delve deeply into the dozen and one trials and tribulations in the shape of insect pests and fungus diseases which can beset the tomato.

In the first place, few of them appear at all with a well-grown ring crop, and, secondly, there are specialist works on the subject which can easily be consulted where necessary.

Nevertheless, certain 'trouble' queries appear in my mailbag each and every season, so brief comments upon the more important may prove useful to readers.

DISTORTED AND MOTTLED FOLIAGE

Usually a symptom of a virus disease infection. Such plants, invariably more or less stunted, may survive to produce some sort of crop but, as the infection is readily spread to healthy plants during the course of training and removing shoots, they are best removed, ring and all, as soon as the abnormality is noticed. If the symptoms appear early in the season, the space left by the removed plant can be filled by training-in a shoot from the lower stem of one or other of the neighbouring plants.

BLOSSOM-END ROT

This is a condition in which the part of the fruit to which the blossom was originally attached develops a hard, black, sunken area. The rest of the fruit is quite edible but the appearance of these black areas must be taken as a sure indication that the water uptake by the roots is—or has been—out of step with the transpiration capacity of the leaves.

In ring culture, blossom-end rot is common with plants which have failed to produce a strong secondary root system in the aggregate by the time the water demand becomes heavy. No disease is involved; the condition is purely physiological and can be prevented in respect of the subsequent fruit only by reverting to watering the rings to an extent sufficient to keep the compost visibly moist at all times. Always providing that the aggregate is not toxic to roots and is kept moist without being water-logged, a return of the ring roots to normal vigour, consequent upon careful watering in the rings, will sometimes result in the formation of a few roots in the aggregate.

Blossom-end rot will then be found to disappear automatically.

BLOTCHY AND 'CORKY' FRUIT

These conditions are, again, physiological and involve no actual disease. Uneven ripening, brown-streaked outer flesh and a prominent white core at the stalk end are all indicative of rank, unbalanced growth and, as such, are much more common in ordinary soil growing than under the controlled circumstances of ring culture. Where these symptoms appear with ring plants, the cause is almost invariably incorrect feeding; too much nitrogen in relation to potash or unduly weak feeds. A return to normal quality can be brought about only gradually and this by changing to straight sulphate of potash feeds for two or three weekly applications, as advised in Chapter 6.

By keeping strictly to 667 liquid fertilizer or the equivalent home-made feed, I find that I am able to avoid these adverse fruit conditions virtually completely even with Moneymaker, a notorious 'blotchy-ripening' variety.

GREENBACK FRUIT

A condition allied to blotchy ripening and which affects some varieties more than others. Thus, Potentate will often develop the typical hard yellowish-green collar round the stalks to a major degree whereas Moneymaker treated identically is free of this uneven ripening symptom. The plants should be treated as for blotchy ripening.

LEAF SCORCH

Specific to ring culture, where some form of ash or clinker has been employed for the aggregate layer, is the occasional appearance of a peculiar type of leaf scorch which commences on the bottom leaves and gradually extends up the plant.

The first symptom is an irregular withering of the leaf margins and leaf tip. After a few days, these areas turn brown and with a gradual inwards spread of the symptoms, the whole leaf may eventually become brown and sere. On one occasion, my own plants were affected in this manner and, after looking carefully into all the various possibilities, I was able to show that this unhappy condition is caused by the roots absorbing some toxic principle—probably a sulphur compound—from insufficiently weathered ash.

Since then, I always test any new source of ash or industrial furnace waste for freedom from toxic principles. The following method is both simple and effective.

Fill a small flowerpot with the ash in question, moisten thoroughly and sow a few quick-germinating seeds, such as of cress. Cover the seeds thinly with further ash, stand the pot on the window-ledge of a warm room and keep the ash damp. If the seeds germinate and have clean white roots when pulled up after a few days, all is well. If there is no germination or the seedlings are stunted and the roots have a burnt appearance, the ash is harmful and should be left to weather in a shallow pile outdoor for at least a further 3 months.

This is the preventative approach: where the symptoms appear in a ring crop as soon as the roots penetrate down to the aggregate, the only thing to do is to lift the rings carefully away from the aggregate, lay down a strip of polythene sheet, apply a 2-in. layer of well-moistened sphagnum peat and replace the rings. As a rule, this results in strong roots being formed in the peat and the plants gradually resume normal growth. The scorched leaves can then be removed. The peat layer must be kept thoroughly moist by daily damping but water-logging, with consequent sourness, must be avoided.

FAILURE TO ROOT INTO THE AGGREGATE

Every season I have reports that a proportion of the plants, as proved when the rings are cleared away at the end of the crop, have never rooted into the aggregate. Obviously, such plants are little better off for water supplies than if they were growing in pots though certainly the ring compost keeps moist better than in a pot due to suction of water from the continually drenched aggregate. Usually the reason for this failure to root-through is over-generous watering in the rings during the first week or two after planting. The roots can then obtain all the water the plants need from the compost so there is no encouragement for further rooting into the aggregate layer. One would expect the roots to go down at a later stage when the demand for water increases and the ring compost dries out more quickly but for some reason—which I have been unable to determine—this does not occur to any effective degree. Either the plants root through by the time the first truss is set or they fail to do so at all. The answer for future crops is to ball-water sparingly at the early stages and so encourage roots to grow into the aggregate in search of more plentiful supplies.

ROLLING OF LEAVES

A peculiar and commonly alarming symptom frequently occurring with outdoor tomatoes, and also found occasionally in the greenhouse crop near the door or beneath ventilators, is a rolling of the individual leaflets inwards to the main vein. In extreme cases, all the mature foliage assumes an almost tubular formation and it is naturally anticipated that the crop will suffer severely.

Leaf rolling is considered to be caused by an excessive flow of sap as compared with loss of moisture by transpiration from the leaves. This circumstance is likely with strong-growing plants where the soil is warm and moist and the atmosphere cool. Another apparent cause is excessive removal of the lower leaves, a procedure which tends to concentrate an undue surge of sap into the remaining foliage.

Fortunately, even severe leaf rolling appears to have no adverse effect on the cropping capacity of the plants though, if caused by excessive defoliation, the defoliation in itself would be detrimental.

CURLING OF HEAD OF PLANTS

When growth in the early season is extremely vigorous, it is common to find the young expanding leaves immediately below the growing point taking on a downward curling about the leaf stalk. This effect is always most noticeable in the early morning after a cool night and is caused by an abnormally high sap pressure. A near-saturated atmosphere and a rooting medium charged with water and rich in nitrogen favour this circumstance which, though in no way harmful in itself, suggests that watering in the rings should be discontinued in favour of high potash feeding with a view to correcting the balance of growth.

Fruit Splitting

With an unduly sharply drained ring compost which dries out considerably between feeds due to poor retention of the feeding solution and limited suction of water from the aggregate, the growth-rate of the plants is uneven. The rate of swelling of the fruit is much reduced in the dry periods and the skin tends to harden and to lose the ability to stretch. Then when, after watering or feeding, the sap pressure increases and swelling of the fruit is resumed, the skin cannot accommodate the increasing size and is split by the pressure within. Such fruit may rot before it is fully ripe and, at best, is spoiled in appearance. Splitting may be prevented by encouraging a strong aggregate root system and by compensating for an unduly open ring compost with half-quantity feeds at 3-day intervals as detailed in Chapter 6.

Botrytis Rot

Botrytis cinerea or 'grey mould' is probably the

widest spread of all the fungi which can cause damage to plants under glass. It grows freely on all types of rotting vegetation and can also penetrate the living tissues of the plant when conditions are suitable to its development.

The name 'cinerea' is descriptive of the dense masses of ashy-grey spores produced by the established fungus and these microscopic 'seeds' in their millions are carried far and wide by the wind. It will be seen, then, that the potential for Botrytis infection is everywhere all the time, and whether our plants are attacked or not depends upon the prevailing atmospheric conditions in the greenhouse. Under stagnant conditions, the air becomes very humid when the plants are filling much of the greenhouse space and, when the temperature falls at night, all surfaces of the plant are covered by a film of moisture—rather like dew on the autumn lawn.

These are the ideal conditions for the Botrytis spores to germinate and for the fungus to grow into damaged spots on the leaves and stems, and into the fruit at the stalk end where the calyx 'blankets' moisture for long periods. Within a few days of infection occurring, dark, rotting areas appear on stem or leaf and the fruit, at all stages of development, drops off due to breakdown of the tissues at the stalk end. The presence of tufts of grey spores is a sure indication that Botrytis is the cause of the trouble.

The answer to this disease is to keep the air moving and the atmosphere buoyant by the freest possible ventilation at all times. From around mid-June onwards —the time when Botrytis begins to be troublesome— my greenhouse has the off-wind roof vent and the door wide open both night and day unless the weather is exceptionally cold. In this way, I avoid Botrytis virtually completely and, at that time in the year, the fruit develops and ripens quite normally under these almost outdoor conditions.

Where the trouble has appeared, all fallen fruit should be collected, infected leaves cut off cleanly close to the main stem and the areas of stem rot cut out with a sharp knife down to obviously uninfected tissue. As surprising as it may seem, the plant will survive if more than half the thickness of the stem is cut away! Wherever cuts are made, the raw surfaces should be protected from further infection by smearing them with a thick cream made by mixing a thiram or captan fungicide with a little water or, failing this, by rubbing the chemist's flowers of sulphur into the cuts.

If the foliage is very lush and dense—which it should not be with correctly-applied ring culture—air circulation can be assisted by cutting out whole leaves here and there, especially where they hang down on to a truss of fruit. Here, again, the cuts should be protected.

Botrytis rot is, however, essentially a trouble which should be prevented rather than cured.

BOTRYTIS FRUIT SPOT

The moist, stagnant conditions which favour Botrytis Rot also cause the appearance of a pale-coloured, lumpy spotting of the fruit. On close examination, these spots, circular in outline when isolated, will be seen to have a dark central speck.

At one time, these spots were thought to be greenfly punctures but work in the 1930's at the since closed Cheshunt Experimental Station showed that the cause is germination of Botrytis spores on the moist skin of the fruit and limited growth in the tissues immediately below the skin. The dark central speck marks the point of original growth of a spore.

These spots seldom, if ever, cause fruit rotting but, being visible even when the fruit ripens, result in a down-grading of quality.

The procedures adopted against Botrytis Rot also prevent Botrytis Spot.

LEAF MOULD OR 'MILDEW'

Olive-brown velvety patches appear on the under-surface of the leaf in late summer and the whole plant may become seared and brown.

Spraying very thoroughly, especially beneath the leaves, with Bordeaux mixture or colloidal copper gives some control but, as with Botrytis, the real answer is to avoid infection by ventilating freely.

All the foregoing are 'plant condition' or fungus troubles. There are also some insect and allied pests which commonly attack the greenhouse tomato whatever the system of culture:

APHIDS OR 'GREENFLY'

These plant lice can cause much distortion of the

young plants and, apart from this direct effect, they are often carriers of virus infections.

Deal with this pest, as soon as it is seen, by spraying with nicotine or derris wash or by fumigating with nicotine shreds.

WHITEFLY

These are tiny moth-like creatures which flutter around when the plants are disturbed. A heavy attack seriously weakens the plants and causes a black, sticky mould to develop. DDT spraying or smoke fumigation with a repeat application a week later gives complete control. Seldom a pest of the outdoor crop.

CATERPILLARS

Tomato moth and various other species at the caterpillar stage can cause havoc to the developing fruits. Here again, DDT in any convenient form provides a 100 per cent control.

RED SPIDER MITE

This minute attacker is neither a spider nor an insect but is none the less destructive for that!

Silvery webs appear beneath the leaves and, indeed, over the whole plant under favourable conditions for the pest. A hot, dry atmosphere and weak, starved plants favour attack so, again, this is not a troublesome pest with well-conducted ring culture.

None of the usual insecticides has any effect on this

pest and control involves very thorough spraying with a white oil preparation, such as VOLCK.

A chapter of troubles? Do not worry, you probably will not encounter most of them but it is as well to be prepared!

11

New Developments

————— ►►❯❚◗◖❰◄◄ —————

Ring culture is firmly established, fully investigated to its ultimate potential and unlikely to be the subject of further advances. However, this system, with its lesser reliance upon soil as the growing medium, has stimulated research along parallel lines and two new methods of growing, dispensing largely or entirely with soil, have emerged of recent years.

STRAW-BALE CULTURE

When compressed straw, such as a standard straw bale, is thoroughly saturated with water and is treated with a source of lime and fertilizer nutrients, it supports vigorous root action. Plants such as tomatoes and cucumbers can be grown effectively on this treated straw by standing the rings on the bale instead of on an aggregate layer. It will be seen that this method of growing differs from ring culture in that the entire root system of the plant is supplied with nutrients. As such, it may be looked upon as a soil-less extension of normal soil growing, especially as soil can, if desired, be dispensed with entirely by substituting fertilized peat for J.I. compost in the rings.

The procedure is as follows: a straw bale is placed on its wide side in a shallow trench lined with polythene sheeting to isolate it from the soil beneath and is gradually brought into a condition of complete saturation by applying 1 gallon of water each day for ten days with a rosed can. Evaporation of the applied water may be largely prevented by keeping the bale surrounded with polythene between waterings and this is particularly desirable where the bale is being prepared outdoors for garden culture.

This gradual saturation procedure is necessary as dry straw tends to shed water and can be saturated only by degrees. When wet, the bale is treated with composting chemicals which promote bacterial fermentation and provide nutrients. The following additions are required per standard bale:

> 1 lb. Nitrochalk
> 1 lb. superphosphate
> 1 lb. nitrate of potash
> 4 oz. magnesium sulphate (Epsom salts)
> 3 oz. sulphate of iron

These chemicals are spread evenly over the surface of the bale and are washed into the straw by repeated light sprinkling.

This treatment promotes intense bacterial activity and the internal temperature of the bale rises rapidly. This rise is particularly marked in a heated greenhouse but it is less apparent in an unheated house or outdoors unless the bale is again enclosed in polythene. Rings filled with compost or fertilized peat, or small mounds of these rooting media, are now placed on the bale and, when the early excessive heat of fermentation has subsided, planting can proceed in the normal manner.

At all times the bale must be kept thoroughly moist: up to 2 gallons of water per day may be required in hot weather when the plants are fully developed. Feeding, commenced a fortnight or so after planting, involves applying 4 oz. of Nitrochalk per bale on three or four occasions at fortnightly intervals and copious weekly liquid feeding both to the rings or planting mounds and to the bale.

The main advantages of straw-bale culture are a warm root-run resulting from the heat of decomposition of the straw and, in the glasshouse, a valuable natural enrichment of the atmosphere in respect of carbon dioxide, a gaseous by-product of this same decomposition.

A more detailed account of straw-bale culture, together with working diagrams, will be found in my companion book, *Tomatoes for Everyone*.

PEAT TROUGH CULTURE

Peat, especially the sphagnum moss type, is widely used for soil conditioning and has been an essential part of container composts ever since Lawrence and Newell evolved the John Innes Composts some forty years ago. Lime-hating plants, i.e. those requiring an acid rooting medium, such as azaleas, have long been grown in pure peat or predominantly peat mixtures, while peat mulching to encourage a surface layer of new roots has been the saving of many a crop of tomatoes failing due to root decay in fungus-infected soil. It seems that we have always known that plants in general will grow well in peat, but that no one had thought of carrying the knowledge to the logical conclusion of peat culture as such.

No one, that is, until of recent years Fisons Ltd., the well-known fertilizer manufacturers, carried out research which led to the introduction of Levington Compost. This is a 100 per cent blended peat physical mixture adjusted in acidity to suit a very wide range of plants and fertilized to achieve vigorous balanced growth until the normal feeding stage is reached.

On a commercial scale, tomato crops are being grown in polythene-lined troughs cut in the glasshouse floor 4 in. or so deep and filled with Levington or comparable peat compost. The same procedure can be, and to a limited extent already has been, adopted in the amateur greenhouse with very good results. The polythene provides a barrier between the healthy peat roots and the stale or infected soil beneath but, since there is no downward drainage, care must be taken not to water the plants to excess or the peat will become waterlogged and the roots will be killed.

An even more recent commercial development is the production of fertilizing formulations which, when mixed into moistened Irish sphagnum peat, produce a complete peat compost as and when required. This convenient variation from purchasing a peat compost as such will doubtless be extended in due course to meet the smaller quantity requirements of the gardener.

It will be apparent to the experimentally-minded gardener that there is considerable scope in the concept of 100 per cent peat composts for trying out their own ideas and theories of growing.

Index

117

Index

Don't seed tomatoe plants too early
Mid April is soon enough.
Keep on a light South facing windowsill
Allow first flower truss to set its
fruit before planting out.